Shed
Couture

Shed
Couture
Peta Mathias

A Passion
for Fashion

RANDOM HOUSE
NEW ZEALAND

RANDOM HOUSE

UK | USA | Canada | Ireland | Australia
India | New Zealand | South Africa | China

Random House is an imprint of the Penguin Random House group of companies, whose
addresses can be found at global.penguinrandomhouse.com.

Penguin
Random House
New Zealand

First published by Penguin Random House New Zealand, 2021

10 9 8 7 6 5 4 3 2 1

Text © Peta Mathias, 2021
Photography not otherwise credited is by Sally Tagg, Emma Bass and assorted friends

The moral right of the author has been asserted.

Design by Cat Taylor © Penguin Random House New Zealand
Cover photograph by Sally Tagg
Prepress by Image Centre Group
Printed and bound in China by RR Donnelley

A catalogue record for this book is available from the National Library of New Zealand.

ISBN 978-0-14-377305-4
eISBN 978-0-14-377306-1

penguin.co.nz

MIX
Paper from
responsible sources
FSC® C144853

Dedicated to my beautiful mother

Contents

Introduction

I have always been deeply submissive to the seductiveness of cloth. I am haunted by the clothes of my past and they often come back to me out of the blue. I will be cooking a meal and suddenly regret selling a Zambesi chocolate-brown netting skirt. Then I wonder what happened to the beautiful dove-grey and lemon top that I wore with it. I remember the feel of these clothes, what I was doing when I wore them, the people I knew at the time, what hairstyle I was rocking, where I was at in my emotional life. Fashion is mysterious, irresistible and alluring.

The kernel for this book came from a series of Facebook and Instagram posts that I started at the end of 2017 that I called Shed Couture. I was going through the shed in my backyard. Yes, I know some people might have gardening equipment, power tools or rusty paint pots in their shed, but I'm not some people. My shed functions as an extra wardrobe for clothes I'm not currently wearing but don't want to get rid of. So, back in 2017, the shed was bulging and I was trying to thin it out, make the contents look like less stuff, wondering, *If I don't remember what's in it, do I really need it?* I admit I chucked or sold some things, but the deeper I delved, dragging out sealed plastic boxes and looking in them, the more I realised I had a shed full of valuable, beautiful clothes and shoes, long forgotten. This is not counting all the clothes and shoes in the actual wardrobes in my two homes in New Zealand and France.

It's fashionable now to follow Marie Kondo and get rid of anything you don't really need or aren't really, really attached to, and I agree with that idea, but if I had done that to its logical conclusion, this book would not exist. I don't need 100 frocks, but I do want them because they are not just frocks — they are beauty, art, history, emotion, memory, identity and, above all, joy.

I constantly and whimsically look at other people's clothes, which, of course, tells me a lot about them. For some people it is about love, self-worth and care; for some it is art and standing out from the crowd; for some it is a utilitarian uniform; and for others just a matter of covering themselves with something, anything. This last group of people look in a mirror and see nothing except that maybe the parting in their hair is crooked. The first group look in a mirror and go, *OMG, thank God I looked in the mirror — how could I have even* thought *of putting those shoes with these pants!*

The way clothes feel against my skin is important, which is why I mostly wear natural fibres of good quality — it's a simple pleasure in this vale of tears we call life. Caring for yourself is not being indulgent and egotistical, and fashion is not a superficial pursuit; it is necessary for self-preservation. Indian women living in the country's slums are beautifully dressed in the best they can afford. They wear ear and toe rings and oil each other's hair, and then, in elegant style, sweep and dust to keep even their humblest lodgings clean.

You should never throw out your most brilliant clothes or shoes because of this word: derivative. There are only so many ideas in the world, and once you've used most of them up you have to go back and

I don't need 100 frocks, but I do want them because they are not just frocks redo the 50s, the 60s, the 70s — even the ghastly 80s. If the enchanting Marni dress you wore to the art opening last week looks like the one your mother wore to Mass in 1956, it's because fashion always come back. So, if your wardrobe feels dated, put the gems in it into your equivalent of a shed and, before long, they'll be the height of fashion again. The only depressing thing about keeping old clothes is you look at them and think: *It is not possible my waist was that small — I've fallen into someone else's wardrobe.* You can't even remember having a waist, let alone the meaning of the words 26 inches. I still have my debutante dress and can still fit into it, but it looks like a first communion dress. My mother's wedding dress looks like it's for a doll.

Chances are, you are wearing clothes right now — could be pyjamas, could be yoga sweats, could be a ball gown. We start getting clothed the minute we are born when the midwife wraps us in a soft cotton blanket. Our mother clutches us in her arms for the first time, and you can bet she's going to pass on a lot more about fashion than the dubious delights of wearing a hospital gown. You might grow up determined to dress nothing like your mother; I grew up with a mother who had a style to envy and who said, 'Always buy good clothes when you're flush so you still look fabulous when you're broke.'

I have sold and given away a lot of clothes over the years, but some were so gorgeous, so expensive and so emotionally linked with an era in my life that I just couldn't part with them. Also, I gained weight — over the period of menopause I put on more than a kilo a year till I weighed 75kg — but I thought that maybe one day I'd lose the weight and wear those clothes again. And that's exactly what happened: after years of patiently believing that being overweight couldn't surely be my lot in life, menopause ended, I started eating normally again and my weight went back to 60kg, where it was supposed to be. I didn't feel ugly when I was fat, but I felt uncomfortable and unfree, both physically and mentally. This is not a story about portion control, it is about the accidental confirmation of fashion sustainability in my life.

So I pulled my old garments from storage in my shed, aired and

steamed them, then put them on and took photographs for social media to show that you don't need to go out and buy more — have a look in the back of your cupboard, storage bunker or shed and you never know what you will find. The posts also showed that well-designed, good-quality clothes don't go out of fashion. I hoped other people would follow suit and put up photos of themselves in their old but ageless outfits, but the only thing that happened was that everyone wanted to buy mine. Drag queens would contact me and say, 'I want your whole wardrobe — how much?'

Alongside my fashionable mother, I was brought up by *Vogue* magazine in slathering thrall to an unattainable, capricious dream that would never be seen in New Zealand anyway. But these days dressing is much more flexible, so the Shed Couture ethos is even more effective. You can wear what you want. For example, there is currently no trendy length for a dress. It seems unbelievable now, but what was deemed a fashionable hem position used to be very rigid, and you wouldn't be seen dead in the previous season's trouser leg width, skirt length, sleeve style, etc. It was a nightmare, and the wrong move would date you immediately. Hypersensitivity over the precise amount of leg to show had been drummed into us at school, where we had to kneel, and if your dress didn't touch the floor you had to let the hem down.

There is still one caveat to fashion: wear clothes that actually suit you because looking good gives you the confidence you might not be feeling. This is what you call asset management. Strangers come up to me all the time — sometimes every day — to tell me I'm beautiful and they love the way I'm dressed. Clearly I'm not beautiful, but I think what they mean is that seeing me makes them happy and they want to share that. I do the same thing if I see someone whose look I like: it makes me happy and I want to verbalise it. The Italian chef Antonio Carluccio once came up to me at a food conference and said, 'I am so happy to see you in this dress — I was with Issey Miyake when he designed it.' He proceeded to explain all the complex motifs on the bright blue 'Pleats Please' dress with the complicated folded collar.

When I was a child and teenager, you would never go up to a stranger and tell them they were beautiful or they looked good — it would have been considered familiar and superficial; in fact, you would never talk to a stranger full stop. Self-confident, brash Americans taught us to compliment each other, just as they taught us to tell our children repeatedly that we love them. No parent in my childhood ever said something as wantonly excessive, intimate and surplus to requirements as 'I love you' to their child. I don't think they said it to each other either. All parties concerned would have just been embarrassed. When I snapped at my six-year-old niece one day she said, outraged, 'You hurt my feelings, Aunty Peta, when you said that.' Feelings . . . we didn't have a feeling till we were 21 — it was illegal. No adult ever considered a child's feelings about any topic whatsoever. I'd never heard of a child with feelings. I'd never heard of a permitted feeling till the first time a boy asked me to marry him. I handled it very badly and laughed nervously. It was like catching someone in the shower when all you were looking for was a toothbrush.

When people make jokes about Anglo-Saxons being repressed, it's not funny to Anglo-Saxons; we think it's normal to keep our feelings close to the chest, in fact we used to keep them hidden under twinset and pearls or business shirts and ties. It was the encounter groups and the 70s that broke open the deep reserve ingrained in us. This era made us talk for the first time in our lives, and, once we got the hang of it, there was no stopping us — we had group therapy and feelings coming out of our wazoos. Now we talk to everyone, including complimenting people on how they look, and we even let our clothes communicate how we feel in a way that was rare amid the uniformity of the past.

Just look at our present prime minister, Jacinda Ardern — she's one of the most admired and emotionally sophisticated people in world leadership today. In 2019 she was included in *Time* magazine's 100 most influential people in the world, in part because of her handling of the racist massacre at the Al Noor Mosque and Linwood Islamic Centre in Christchurch. *And* she's good looking *and* she dresses well.

She communicates so much by showcasing garments by New Zealand fashion designers, such as Maaike, Ingrid Starnes and Alex Dixon, not to mention wearing the stunning Māori kākahu when she visited the Queen, and the hijab to show sympathy with the victims of the massacre. She knows the power of clothing and how it can communicate positive messages. In 2019 she was put on the cover of the September British *Vogue* by guest editor Meghan Markle. Jacinda, along with others, was chosen as part of a varied selection of women who are each working for change, equality, justice, open mindedness and, of course, kindness. The New Zealand garments Jacinda championed in the feature included items by Juliette Hogan, Andrea Moore and Allbirds shoes. Confidence, feelings, communication, politics: clothing is more than a few threads thrown together for warmth or modesty, it is the fabric of deep philosophical considerations.

I can't imagine a world without beautiful garments — it would be like being a singer without a song, a cook without knives. I love wearing beautiful clothes for the same reason I love cooking and good food — it's an easy way to make myself happy. With both I am nourishing myself, living life to its fullest and hopefully saving the planet a little at the same time. Just as you should cook with love, you should dress with love. Just as you first eat with your eyes, so you dress with your eyes. Just as a chocolate and almond cake looks more tempting with drizzled chocolate, flowers and slithered almonds on top, a salad is perfectly fine on its own but shimmers and says 'eat me' when it's dressed with shining olive oil and fresh lemon juice, and chips are just chips until you put a dollop of aioli on top, so a black dress is just a black dress till you put a neon yellow Pleats Please scarf with it, and a handbag is just something to put your lipstick in unless it's a handmade Hermès Birkin bag (even better if recycled). With both food and clothing, there are constant tribal decisions to make — kitten heels or jandals, poached or steamed, linen or silk, carb or fat, minimalist chic or grunge?

Is my fashion and food philosophy fanciful? Well, both are subject to trends, seasons and clever marketing (some of which we'd like to

I can't imagine a world without beautiful garments — it would be like being a singer without a song, a cook without knives.

forget — remember power shoulders; remember cronuts?). Food and fashion both require inventiveness and self-expression. There are lots of fashion houses with restaurants in them, and fashion/food magazines, but I am surprised there are not more — it seems like such an obvious match. They both give structure to your day: breakfast and block-printed Indian pyjamas, lunch and Marni slacks, dinner and an evening gown. I mean, *hello*.

Many designers are very good cooks; for example, the Tunisian Azzedine Alaïa. In my research I found a wonderful woman called Elettra Wiedemann, daughter of Isabella Rossellini and granddaughter of Ingrid Bergman. She is a model *and* a food writer, and also an activist on health and sustainability. Her very modern approach to fashion and food has turned her into a poster girl for good living. She now lives and works with her partner, son and mother on her mother's organic farm on Long Island, where she makes her own bread, grows vegetables and wears elegant, simple clothes.

It makes sense, therefore, that many designers have put food prints all over their designs. I found my Marni eggplant dress at Scotties recycled selection in 2007. It was a little tight, so I painstakingly let out all the darts. It has a gorgeous pattern of what look like purple aubergines on a teal background, and has a plunging back neckline. I wear this dress often with chunky Marni earrings and platform shoes.

It's always been the case that haute-couture designer ideas, like huge ruffle sleeves, will eventually trickle down to the high street, just as fabulous French *macarons* are now made by unfabulous McDonald's. Both ruffle sleeves and the *macaron* sing a song we can't resist and they connect us to history, beauty, art and tradition, so even if there's now a watered-down version, that doesn't mean you should ban them from your wardrobe or pantry.

Another connection between fine food and beautiful clothes is that both require quite a bit of assiduity and a conscious desire to make them an integral part of your life. You don't have to be rich to do so anymore, but you do have to go to a bit of trouble to buy good food and

Food and fashion are the perfect recipe for my eggplant dress by Marni.

take the time to use an inspiring recipe before it adds to your quality of life, and it's the same with clothes. There are two kinds of fashion: the first, runway or haute couture, is affordable only to Saudi princesses; the second is street or off-the-peg fashion, which the rest of us wear. But you can preen with the princesses by finding the recycle shops that stock second-hand haute couture (and there are more of them around than ever before), by scouring the good Instagram feeds, and by tumbling out of bed in the morning and deliberately putting something beautiful on. When times are tough, I have to say that good food usually wins out because a dinner at Le Taillevent in Paris is going to cost a lot less than a Dior jacket. But, of course, a Dior jacket is going to last a lot longer.

I once wrote that because food makes us happy, we fall into the trap of assuming more food must mean more happiness. The same can be said about clothes: buying more will only make you happy until you get back from the shop. Buying better less often or making the most of what you already have are really the best ways to sartorial satisfaction.

The point of this book is to explore my ideas for sartorial satisfaction, to tell the stories associated with my love of clothes and fashion, and to share the message of accidental sustainability.

Recycled Marni earrings and necklace teamed
with Pleats Please's timeless style (© Neil Gussey
& Verve Magazine).

CHAPTER 1

Slow Fashion and Sustainability: make well, buy well, re-sell

Interestingly, Coco Chanel was the first designer to say 'less is more', which she advocated partially for reasons of sustainability but mostly as a question of taste. She found it incredibly vulgar to be overindulgent. When she first started, her clothes were so different from the fashion at the time: for a start, she liberated women from the corset and made beautifully designed clothes affordable, practical and accessible. 'Elegance is refusal,' she always said. What a woman. And she's right: a few really good clothes are a much better investment than lots of bad-quality ones. Remember Scarlett O'Hara making that stupendous green velvet gown from her lounge room curtains! Even the Duchess of Cambridge makes a point of wearing the same outfits more than once, as does her husband Prince William. As does Michelle Obama. I am an accidental sustainability icon and maybe so are you.

The problem of excessive waste is recent. My parents and most people of their generation lived a sustainable life: they had no debt apart from a mortgage, bought nothing on 'time payment' — if they needed a washing machine they waited till they had the money to buy it, and Mum made all her own and her six children's clothes. Nothing was ever thrown out. Socks and torn clothes were mended. Clothes were passed down to other children (the youngest probably not being too thrilled with that). We cherished clothes as we cherished our relationships — respect for clothes is like respect for people. With the advent of Donald Trump, the world seems to have learned that it's okay to be mean, rude and careless with each other, which reflects onto everything: eating, dressing, living, loving. We did not eat junk or fast food except for fish and chips in my childhood (it didn't exist when I was born in 1949). My parents grew vegetables, fruit and flowers and had chickens in the backyard. This was common for city folk — we didn't live in the country. If something broke, you fixed it. I took my toaster to a repair shop recently and the nice man laughed at me.

It may be difficult to recognise that your love of fashion is environmentally damaging. The fashion industry is one of the biggest and most influential in the world, generating around €1.5 trillion annually, and while we currently have 7.8 billion people on the planet, around 150 billion new garments are manufactured a year. The world (including poor countries) is simply producing far too many clothes. A lot are worn only once and a lot are not biodegradable. Textile production is responsible for 10 per cent of global greenhouse-gas emissions and for almost 20 per cent of industrial water contamination. Fashion, especially what's called 'fast' fashion, which focuses on cheap prices and high turnover, is one of the worst polluters globally. Americans alone toss out 14 million tons of clothing per annum; the whole world wastes 92 million tons! That's around 80 pounds or 35 kilos a person. Here, the amount of textiles sent to one Wellington dump has doubled in the past 10 years. In Auckland, textile waste makes up 4 per cent of landfills, and that's expected to rise to 6 per cent by 2040. If you keep and wear a piece of clothing just nine months longer than planned, you will reduce the carbon, water and waste footprint by 30 per cent.

Accounts of wastage get worse. In 2017 British luxury brand Burberry *burned* £28.6 million pounds worth of unsold products. Lots of other brands burn unsold products too. Even H&M, who do have an ongoing sustainability programme to close the fashion loop, burn 12 tons of unsold garments a year. We used to think that donating to charities was a good way to get rid of unwanted, cheap clothes but this no longer works so well, especially if the clothes are bad quality and they sit in huge warehouses unused or are dumped in landfills.

If we stop buying crap clothes, then companies will stop making them. It's not just rubbish quality and dreadful waste we are concerned about, more than ever it's concern for the life and working conditions of the person who is actually sewing that one-dollar T-shirt. We all know about the dreadful incidents in Indian and Bangladeshi sweatshops, including the deadly collapse of the Rana Plaza clothing factory in 2013. Tearfund New Zealand issues an annual ethical fashion report highlighting both the ethical and environmental issues of fast fashion: you can check out which companies are scoring well and which really need to commit to the six better practices they pushed for in 2020. Meanwhile the documentary *The True Cost* details the impact that the fashion industry has on our world: 'The price of clothing has been decreasing for decades, while the human and environmental costs have grown dramatically. *The True Cost* is a ground-breaking documentary film that pulls back the curtain on the untold story and asks us to consider, who really pays the price for our clothing?'

Aneeth Arora's Péro line combines international style and Indian handcraft.

If you are buying cheap there is a reason for it. Being thrilled with yourself for buying that bargain T-shirt is indulging in a false economy, much like buying KFC — it's a poor source of protein and you'll be hungry again in an hour. We need to consider who made this T-shirt, how much they were paid (about one cent, I'm guessing), what their working conditions are, and whether they belong to a union. In order for clothes to be a joy for *everyone* in the supply chain, we have to be less about endless relentless growth and more about prosperity and abundance for all. Lots of designers of not only clothing but all sorts of products now have Cradle to Cradle certification. The concept of Cradle to Cradle is to ensure that anything we create can completely return to the earth at the end of its life, or it can be used for something else.

Circularity involves creating a no-waste, reusable, regenerating circle right from the moment the garment is designed. That includes a green supply chain, eco-friendly textile consciousness and consumer activism. Circularity also includes upcycling, resale and rental. Think global, wear local seems to be the mantra. Also buy less, buy better, as this positively impacts the wearer, the maker and the environment. A slowed production rate will unlock the potential for future economic development. The term 'slow fashion' was first formulated by Kate Fletcher of the Centre for Sustainable Fashion at the University of the Arts, London, when she identified the need for a slower pace in the industry. It is a concept, an awareness, a determination for the future that encompasses the ideals of sustainability, circularity, fair working conditions and good quality, but interestingly there is no official slow fashion organisation like there is for the slow food movement.

The idea and value of handmade clothes has a perception problem, though. When I had silk/cotton scarves designed in Auckland and hand-printed with specially made blocks in India, a design that would never ever be seen anywhere else, some buyers questioned how they could possibly be so expensive and said they could go down the road and get something similar for half the price. It's made in India, they said, so it must be cheap — right? There was a further issue. A handmade fabric and garment can never be perfect — that is the nature of handmade. Humans are inconsistent and create beautiful variations; machines don't. Which would you prefer to eat: a handmade burger that will vary each time it is made, depending on what ingredients are in season and the mood of the maker, or one pumped out by a factory, always tasting the same? Our need to have uniformity and perfection is a mistake, and the individuality of artisans (and India is the biggest producer of traditional handicraft) must be understood, respected and supported. The inconsistency of artisanal clothes is a value, not a flaw. Buy one good scarf, not five cheap ones.

Things *are* shifting, however. There is definitely more awareness about supply chains and sustainability, both with clothing companies and the public. For example, Dutch jean makers Denham have joined up with Candiani Denim, who are big on responsible fashion, to make the world's first biodegradable stretch denim made from plant-based fabric, with the stretchiness coming from natural rubber. Northern countries like Finland and Sweden are leading the way with the 2020 Helsinki Fashion Week inaugurating the first international sustainable designer residency programme. The designers worked together on transparency and reusing materials to create their collections and were filmed and live-streamed each day.

Global Fashion Agenda, which encourages the industry to get their act together on sustainability, is behind the wonderful annual Copenhagen Fashion Summit. Here they nut out education, insight, priorities, incentives and, most importantly, solutions. To counter the 4 million tons of unsold clothing discarded in Europe every year, France has passed laws forcing stores to give the clothes to charities and to those in need. President Macron estimates that this change will cut carbon emissions by 250 thousand tons. France is also trying to convince other countries to establish a sustainable code practice. Meanwhile, designer Stella McCartney is *the* queen of actually doing something concrete — since 2013 she has been producing sustainable clothes: no fur, no leather, no PVC, no forest-destroying rayon, no cashmere. She is really the heroine of mindful fashion. My friend Greta in New York introduced me to another sustainable designer, Gabriela Hearst, who says 'waste is a design flaw'. So, times are a-changing.

A word on cotton. Cotton is famous for needing lots of water to produce. Originally cotton crops didn't require much water at all (like an olive tree they like struggling in hopeless soil), but now, because of hideously unsustainable farming methods, the plantings are tripled. Monsanto then came along and severe spraying started. As the World Wildlife Fund says, 'current cotton production methods are environmentally unsustainable', and we have to contend with carcinogens, pollutants and the need for lots more water. The old-style organic, sustainable cotton India has now started to produce again is much more costly, but, as I constantly preach, buy less and buy better.

Whereas buying second-hand clothes from charity shops used to have a stigma, now it is seen as a desirable and intelligent thing to do — it's termed upcycling or thrift-flipping. I have a friend in France who shops exclusively at a charity shop and she wears almost entirely designer clothes. A lot of the designer clothes I have bought over the years have been from designer recycle shops — I mean, I'm talking

A Christian Lacroix scarf that I've had for many years; it's great for adding a splash of colour.

fabulous Marni, Dries van Noten, Christian Lacroix, JP Gaultier clothes in perfect condition — and if I feel like it, one day I will pass that beautifully made dress on to someone else. (By the way, one of the best recycle designer online shops in the world is The RealReal.) It is now predicted that the second-hand clothing resale market (both mass-market and luxury) will overtake fast fashion and, in fact, fashion in general if it keeps expanding at the present rate.

Another really good idea is renting fashion. Panoply in Paris was started by Ingrid Brochard and Emmanuelle Brizay, who figured out that a lot of women would LOVE to wear high fashion but could never afford it in a million years, or just felt like they still had 'nothing to wear' despite a wardrobe bursting full of clothes. They came up with the concept of renting clothes, having a special night out (or three), then giving them back. You, too, can wear the clothes that your favourite stars, fashion icons and designers wear — off you go wearing the latest Sonia Rykel, Marc Jacobs, Chloé or Courrèges. You can rent the whole look: dress, shoes, bag and jewellery. You order online or you can go into the showroom, the clothes are delivered to you and then picked up, and cleaning is included in the price. There are now similar

Some of the clothes in my shed are 30 years old, and they don't look dated.

renting businesses all over the world, such as Designer Wardrobe in New Zealand.

When Lidewij Edelkoort, the fashion trend forecaster (from Trend Union), says 'Fashion is dead. Long live clothing', she could well be right. Dutch born, the globally influential Li is the dean of hybrid design studies for the prominent New York fashion college Parsons School of Design. For the past 40 years she has never got her predictions wrong because, in her words, she just keeps her ears and eyes open to what clothing ideas are circling and trusts her intuition. In the late 1980s, for example, she noticed New York City messengers wearing leggings and predicted it would become a trend. Did it ever: a billion-dollar trend. Even I have made tiny predictions myself. In my 2008 book *Can We Help It If We're Fabulous?*, I wrote 'the name of the game is that there is no more name — every woman wears what she wants, everything is in and everything is out. Nobody has the cultural authority to dictate fashion. The days of rules are gone.' This has turned out to be true and remains the case. Some of the clothes in my shed are 30 years old, and they don't look dated. In an interview with *The Guardian*, Li has said 'I can work 25 years ahead.' She says that in 2000 she predicted that by 2025 there will be a 'fusion of everything contrasting . . . Work and leisure, man and woman, very intense sport, but also laziness and meditation.' And a new word has indeed entered the fashion lexicon — gender-fluid: a liberating mix of both male and female, or neither.

Li says the closing of the famous New York department store Barneys in 2020 is proof that people are sick of 'fashion as usual' and sick of overconsumption — they want change. She suggests designers 'make much less, make it better and make it more expensive'. She reckons there's a shift happening to determine success by levels of happiness rather than dollars. Reflecting that shift, Li believes this decade will focus on 'proven, functional, workwear shapes', but along with a romantic look, high waistlines and volume: 'Eccentric clothes, romantic clothes.'

When Covid-19 hit, Li said in a *Business of Fashion* podcast that the virus could be seen as 'a representation of our conscience . . . [bringing] to light what is so terribly wrong with society, and every day'. Her predictions for post-Covid are that it will 'completely reset the way we produce, dress and consume', designers will make fewer collections every year and the 'breakdown of the system could be a source for good in the long term'. She loves the idea of loosening the stranglehold of the pre-Covid set-up, paving the way for 'the age of the amateur'. In an interview in the *New York Times* she said, 'We will come out of this, like we come out of a war. The buildings are still there, but everything is in ruins. We will want two things: security and to dance. We will be aching for something new, to refresh our personalities.' I love Li Edelkoort — she is the revolutionary thinker who has the most resonance for me.

According to Clare Press of Wardrobe Crisis, fashion schools are brimming with eco-warriors and brilliant students who are eager to do fashion differently. As usual, London — long famous for producing creative, earth-shattering designers — is leading the way in such art and design colleges as Central St Martins and the London College of Fashion. Their young graduates, Clare says, are asking questions: Is the way we do things still valid? Do I still want to be part of it? How can I prevent waste and redesign the system?

During lockdown in April 2020, Fashion Revolution, a collective of fashion people from all over the world, including designers, academics, writers, business leaders, policymakers, brands, retailers, marketers, producers, makers, journalists, workers and fashion lovers, had their annual week of conferences, panels, workshops and discussions — online this time. This British-based global movement was founded by Carry Somers and Orsola de Castro in 2013 to commemorate the 1134 people (mostly young women) who died and the 2500 who were injured in the collapse of the Rana Plaza garment factory in Bangladesh. They believe that no one should die for fashion, and encourage people to ask the question 'Who made my clothes?', with the hashtag #whomademyclothes appearing all over social media. They encourage people to have a really good look in their wardrobes and think about things like what the clothes are made of, who made them and in what kind of conditions, what could be looked after better, and what could be worn more or perhaps swapped with someone.

When you swap clothes, you bond with your friends, colleagues and neighbours, you extend the life of the garment (which you may not have worn for years), reduce the carbon footprint, and buy and discard fewer clothes. Another thing you can do when looking deep into your wardrobe or shed is to fall in love again with the valuable clothes you have stopped wearing by changing them — take the hem up, remove the shoulder pads, restyle, let out, take in, combine them with other clothes in a way you would never have done before. Fashion is a tool of rebellion, so you can express yourself any way you want. Think hippies, Vivienne Westwood, Madonna wearing her undies on the outside, goths — all rethought the traditional in startling ways.

My accidental sustainability stood me in good stead during the lockdown. In my backyard shed I found lots of designer jumpers, pyjamas and boots I never wear because I never live in winter. Around 1995, I made a decision to spend half my life in New Zealand and half in France. I started off spending one month in France, the next year spending two, and so on till I got to half and half, ensuring a life of endless summers. After I was stranded in New Zealand because of Covid, my friends thought I was going to have to wear 37 layers of summer clothes to keep warm. *Mais non!* I was very warm in my Jean Paul Gaultier, Rodier, Dolce & Gabbana, Hermès and Stefanel jumpers and the winter coats from my 1980s Paris days, not to mention dozens of Indian shawls I had hardly ever worn before. Once retrieved from my shed, ALL were in perfect condition, as if time had stopped. Now I know what I will look like in the care home when I am a hundred: a wreck with white roots wearing Prada. Making a statement is all I ask.

I also realised in lockdown that I was able to not spend money, and I think that might just become a habit. What I did miss was physical contact, especially with my baby great-nephew — I had to live without kisses and cuddles and the feel and smell of his soft skin. While designer fabrics are pretty close, even they can't quite compare.

Do stripes ever go out of fashion?
Like mother like daughter in the 60s.

CHAPTER 2

My Story in Clothes

The sacred purpose of clothing is to tell a story: the story of you. Your choice of colour, fit, fabric and style all convey subtle messages about who you are, what you do, how you feel about yourself, how you feel about the world. They show your deepest complexes and highest self-esteem. A woman who wears jeans and a shapeless T-shirt day after day is saying, 'Hello. I don't want you to notice me because I have low self-esteem and hidden beauty, and I'd like it to stay hidden, if you don't mind.' A woman who wears a tight red wrap-dress and high heels is saying, 'Hello. Go ahead, honey, have a good look — what you see is what you get.' A woman dressed in Laura Ashley is saying, 'Hello, I'm Pollyanna. I'd prefer you to think I produced my children without having sex and I only eat organic food.'

A woman poured into skinny, studded, leather jeans, black T-shirt, nose studs, powder-blue Doc Martens and a cowboy hat is saying, 'Hello. I'm a diesel radical with a fun streak and a weakness for straight women.' A woman wearing Issey Miyake Pleats Please is saying, 'Hello. I'm a stylish intellectual, darling, who should have gone to art school. I wear Miyake because I travel endlessly, don't iron and am probably quite fabulous.'

Uniforms

We all get our first ideas about fashion from our family, then at school and perhaps church, a lot of them revolving around uniforms. My two sisters and I turned up to Mass on Sunday in identical outfits with starched petticoats — they thought it was annoying, I thought it was fabulous (I mustn't have been turned off the idea of uniforms at that point). I don't think my three brothers were subjected to this triple dressing technique.

Uniforms tell a very strong story, most of it being: don't fuck with me under any circumstances. They were invented for various reasons — to save your brain power for a loftier goal than fashion, to show you are serious about your job, to deter unwanted interest, to eradicate individuality, to minimalise class difference, to strangle creativity, to control and discipline. Inevitably, given all that, certain uniforms have been subverted. Take the French maid's uniform, which was originally designed to subordinate women as drudges but subsequently as objects of sexual titillation. In *The Handmaid's Tale*, Margaret Atwood explores the way uniforms can be misused: the bright-red uniform is used to limit and constrain and signify that the women are the reproductive organs of the new country. The large white bonnets cover their hair and obstruct their vision (of the outside world). It is indisputable that uniforms have hugely ambiguous associations with sexual impulses and moral behaviour. And then there's the fetishising of the military look, but, aside from the kinky, details such as the brass buttons, braids, medals, frogging (buttons and loops), epaulettes, over-the-top collars, free use of gold and silver, jumpsuits and camouflage have moved into the civilian uniforms of such professions as pilots and other transport operators, hotel receptionists and porters. They have also crept into fashion wear, from Yves Saint Laurent's 1960s pea coats, to Freddie Mercury's braided jacket, to the recurrence on the cat walk of jumpsuits and aviator sunglasses.

With every uniform, there are little ways of getting around the homogeneity to let people know you are not an android: ask any schoolkid who has worn a tie about the different ways they can be knotted to make a stand. But an overtly skinny or overtly fat knot can then become yet another uniform style for a specific clique. Even without uniforms, our choice of clothes can signal that we belong to a certain group or tribe, revealing our age, class, status, job, gender, culture or political leaning (for instance, the red or the blue of the main political parties). It's so interesting how we use clothing to tell a story and put ourselves in a box. Of course, following fashion slavishly can also be a type of uniform: rather than looking just like the latest mannequin on a magazine cover, dressing with flair should make you and your unique personality shine.

Uniforms can be haute couture. Take flight attendant uniforms, which are supposed to signify the brand of the airline, as well as the sovereignty of the country they are representing, plus be easily recognisable by the passengers. You might think they should be all about practicality, but in fact the classier the uniform, the more seriously the airline is taken. That's why a whole raft of top-end designers have turned their hand to these uniforms: Vivienne Westwood designs for Virgin Atlantic, Christian Lacroix for Air France, Pierre Balmain for Singapore Airlines, Yves Saint Laurent for Qantas, Zac Posen for Delta Airlines, and Trelise Cooper for Air New Zealand. The wildest and most outrageous were the Emilio Pucci uniforms designed for the now defunct Braniff International Airways in 1966: colourful geometric prints in leggings, boots, suits, culottes, tunics and hats. What started off as a very conservative uniform has now become associated with haute couture.

The dubious joys of uniforms and lisle stockings.

From black to white, and more silly headwear.

Which brings me to the opposite extreme: the habit. I first wrote about my experience with the nuns of my childhood schooling in my book on Ireland, *Burnt Barley*. Those nuns had such a strong influence that I wrote more about them in *Can We Help It If We're Fabulous?*, where I explained that they were such a source of mystery to my five-year-old self that I was determined to understand their fashion sense. As far as I was concerned, anyone who wore such heavy layers of clothes through the heat of summer had to have something to hide. After all, there was no secrecy about my mother's summer clothes: she wore a bra, knickers and colourful sun dresses. She was very pretty, had short dark hair and wore pink lipstick. But the nun's all-concealing outfit of so much black — stockings, shoes, long habit, veil — topped by the magpie contrast of the white headgear required explanation. So I peeked under a passing habit to see what on earth they could be hiding . . . and what I discovered was a revelation: the sensation of a strap on my tender skin.

Our uniform at primary school was likewise almost all black — heavy shoes, thick socks, pleated serge uniform, hat and cardie — with just a stripe of yellow in the black tie. They were dressing us as mini nuns, warding off any backsliding, heathenism, individuality, joy, brightness, will to live . . . No wonder I don't wear black now.

At secondary school, I was very relieved that our uniform was a much more lovely blue. Being the colour associated with the Madonna it was deemed virginal, so I should have realised a new level of control was about to be imposed: the biggest concern to the nuns now was what went on under our uniforms. Even though we wore suspender belts, they had to be white, because the black of our primary school outerwear was now regarded as sinful if worn underneath. Go figure. I did ask if I could have black underwear, but was told, 'Don't be a dirty girl.' But at least I was a day girl; the poor boarders had their underwear policed much more stringently. Bikini briefs, which had just come into fashion in the 1960s, were regarded as even more depraved than black undies and were incinerated: it was compulsory to wear underpants to the waist. The nuns would probably have preferred them to be worn to the armpits. Fortunately, after protests, girls no longer had to undress and bathe in a cotton robe so as not to be tempted by their bodies. But our lisle stockings, made from polished cotton, were tripe-coloured and bulletproof, and the bras had been designed to survive a nuclear explosion. That any boy could get past all this, let alone managed to impregnate anyone, remains a scientific mystery to this day.

The next ridiculous uniform I had to be encased in was my nurse's kit. When you had been brought up to wear outlandish costumes, indulge in weird ceremonies and lead a semi-military life, it was nothing to slip into the iron lung called a nurse's uniform. The interesting thing about this particular uniform that differentiated itself from others was that it was feminised to deny the woman professional power — it was a cross between a nun's habit and Coco Chanel. In fact, the original nurse's uniform was indeed derived from the nun's habit. After all, nuns originally looked after sick people before Florence Nightingale arrived in the mid-nineteenth century. Her colleague invented the first nurse's uniform, which was a blue dress, white apron and white cap. By the 1990s, American nurses refused to wear these uniforms, and now most wear scrubs, which could never in a million years be thought of as titillating or feminine or haute couture.

History is littered with men who have fainted at the idea of having sex with a nurse on the operating table (okay, maybe the matron's desk) — there's something about the restrictiveness of the uniform that was arousing to just about everyone except the person who had to wear it. The white *severely* starched dress that crossed over at the front was held together with removable buttons. They got caught in the industrial washing machines, so you had to attach them every time you put on a fresh uniform. How they got all the blood, guts, faeces and pus out of those garments, I don't even want to guess. The white belt was wide and uncompromisingly rigid, the lisle stockings thick, the winged cap rigid, and you could actually hear us as we moved as all those fabrics rubbed. The only saving grace was the glorious royal blue woollen cape, reminding us that we truly were angels of the Virgin Mary. We weren't allowed any form of self-expression whatsoever — no earrings, rings, decorations, different shoes, no excessive make-up. We wore a fob watch pinned to the chest. My first act of subversion was to wear a blood-red petticoat under my white nurse's uniform. This was considered by the matron to be just one slippery step away from having sex with a python.

The last uniform I was ever to submit to was the chef's uniform, invented in the mid-nineteenth century by famous French chef Marie-Antoine Carême. Whereas with the nurse's uniform I was made to understand my career choice was just short of sainthood, with the chef's one I felt very cool and rock and roll because it was in the 1980s that chefs were starting to be seen as superstars rather than drudges who were one step above drainlayers. It was also groovy and unusual for a woman to be a professional chef in Paris in 1980 because the scene was dominated by males. A female cook had to open her own restaurant if she wanted to make a reputation, as she would rarely be given a chance to get to the top under a male chef. In the kitchens I worked in for 10 years in France, I wore a white T-shirt or chef's jacket with the knotted cloth buttons, black-and-white check pants, long white apron and clogs. I refused to wear the toque (chef's hat), the design of which, incidentally, dates back to the sixteenth century. Instead I tied my long black hair on top of my head.

I ended up being so brutalised by my glamorous but hard life as a chef that when I teach cooking classes these days I decline to wear a uniform. To me, the chef's jacket now symbolises a mix of exhilaration, exhaustion, excitement and madness. I also get the pip when cooking teachers, who have never been chefs, never worked in a professional kitchen and never done the hard yards, wear chef's clothes. They are not chefs — they are cooks in chef's clothing. In my cooking school in Uzès in the South of France, my kitchen is full of colour, and I joyfully wear the most beautiful clothes I can find in my wardrobe, together with stylish shoes and lots of make-up. I feel like a man who has finally thrown away his suit and tie.

Even as a baby, I was happy to pose in my finery, here showing off my beautifully smocked dress.

DIY

I have always sewn, knitted, stitched, made, embroidered, smocked and repaired, and so did my mother and so does my middle sister Keriann. Mum and the sewing classes at school made sure that I could make something out of anything; a flour sack should the occasion arise. And I needed to, as my enthusiasm for clothes was boundless. Even as a little child I changed outfits several times a day till I was happy. I was so in love with clothes that, as a 10-year-old, my dresses, jumpers, blouses, pants, hats, gloves and shoes were all colour coded and neatly folded and ordered. No one made me do this. If a passing psychiatrist had looked in my wardrobe, they would have kindly suggested a session on the couch.

I wore smocked dresses as a toddler, then grew up to make smocked dresses from Liberty print and Viyella for my nieces. Not that smocking is only for baby clothes: I even smocked adult dresses and nighties for myself. In the 1990s I bought a cream silk/wool mix Akira jacket. It is very voluminous and entirely gathered with rows and rows of running stitches like the back of a smocked garment. At the back and on the sleeves the designer left the threads to hang long and loose, giving the garment a breathtaking sensuality.

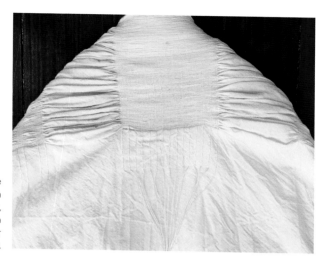

Haute-couture smocking on my Akira jacket, still holding firm after a quarter of a century.

I can't remember if my mother or grandmother smocked my baby dresses or not. I learned the technique from my friend Lesley, who was smocking women's clothes for designer shops in Auckland. Being nimble of finger, I was enchanted by this pretty but practical form of embroidery. The thing with smocking is that it is not just for ornamentation, as is handmade lace, for example. The beauty called lace serves only one purpose and that is of decoration — it doesn't keep you warm, doesn't hide your body, is fragile and expensive. It was a luxury product requiring great skill to make, with a complicated arrangement of bobbins and needles, and it showed that the wearer had wealth, status and taste. Going back at least to the twelfth century, the old English tradition of smocking doesn't have such a high-class origin, being mostly done for peasants by peasants, and it had a practical purpose in being the original form of elastic. The gathered pleats were embroidered to allow them to stretch, resulting in a very useful garment that would slip over your head without the necessity of buttons or ties. Where lace thread was made from silk and linen (and sometimes gold and silver), smocking thread was normally cheaper cotton and rough linen.

Originally smocks were protective garments worn over clothes. They

A selection of smocking sewn by my own fair hand.

had big embroidered collars that indicated the person's job, and were made of linen treated with linseed oil to make them waterproof, given the outdoor nature of a peasant's work. The industrial revolution ended the wearing of smocks because they were too cumbersome for machine work. Folk still wore them for special occasions, and in time well-off ladies started wearing glamorous versions, given that by then they had become associated with artists and nostalgia rather than menial work.

The width of the fabric has to be three times the finished product. First you have to make the pleats by gathering them together with threads, which are fixed at the end of each row. Next you turn the fabric over and embroider the tops of the pleats together with decorative stitches. Then you turn the fabric back over, pull all the retaining threads out from the back, and you are left with a beautiful decorated stretchable front. Originally smocking or gauging was done by eye and the stitchers were clever enough to gather and embroider at the same time. I used to buy iron-on dots and would use them as a guide for the gathering with running stitches. If you choose a striped, dotted or gingham fabric, of course your guideline is already there. These days you can get little pleater machines, but I never took to them as the dots were uneven. There are many stitches

like cable, flowerette, wave, honeycomb, trellis and bullion. The fabrics I mostly used were silk and cotton in Liberty and Souleiado prints.

After I graduated and left nursing behind, I was gainfully employed sewing beautiful hippy clothes for the Cook Street Market in Auckland. These garments were just like the ones I made for myself. Selling the pin-tucked, white-cotton and Liberty-print, ribbon-bedecked dresses I produced at home from my Singer sewing machine was much more fun than irrigating people's kidneys at Auckland Hospital. What goes around comes around, and dresses like the ones I made for the market are now fashionable again.

As already mentioned, my mother made all her own clothes and ours when we were children. She and I made my doll's clothes, too. When she was a young single woman in Sydney, she would buy fabric for a ball dress, lay it on the floor and cut a pattern freestyle, sew it up on her Singer (invented in 1851) and wear it to the ball that night. In my childhood, during the evenings, she and Dad would roll skeins of knitting wool, talk, learn dance steps and play the piano. Television had yet to be invented so, quaintly, people listened to the radio or talked to each other, while their hands created.

Mum was a girl from an Irish family of nine children. She won an important grant to study art when she was in secondary school, a grant her father refused to let her accept as she was only a female and what a waste. She and I (when we weren't getting on each other's nerves or vying for control of the universe) would spend ecstatic hours poring over patterns, choosing fabric, pinning, hemming, tucking, measuring and ironing. I still love ironing; if only you could do that with life's problems. That's why I like the delete button on Facebook — it saves me from deleting people in real life. I still love mending for the same reason I like bread-making: it's calming. You are forced to do something slowly, you go into your own secret world; the feel of the fabric and feel of the dough give you pleasure way beyond the prosaic activity. The pleasure of smocking, embroidering, knitting, etc. is not only in creating something beautiful but in doing something slow and deliberate, which is calming,

I still love mending
for the same reason
I like bread-making:
it's calming.

meditative and leads you to think you are a domestic goddess. And just as traditional recipes are the best way to access another culture, dressing is the thread that connects society. Women embroider in groups so they can stitch and bitch together.

My paternal grandmother, Jessie, was an elegant woman of perfect posture and impeccable taste. She lived with my grandfather in a rather grand house in Remuera and, in the interests of my mother and I not strangling each other, I spent as much time in her lovely company as possible. She was tall, slim, ladylike and always beautifully dressed in overseas designs she picked up on her frequent holidays abroad. Their home was civilised and rather formal, and I worshipped the ground she walked on. I liked everything about her — her blood red lipstick, her permed hair, her silk stockings and underwear, her furs and the fact that I could make her laugh till she cried just by being myself. Grandma and I would get all dressed up in our best shoes and gloves and everything, jump in the Humber Hawk and go shopping in posh department stores. They were drapers by trade (my great-grandparents owned Le Bon Marché department store in Cardigan, Wales) so I like to think I partly inherited my love of fabric, crafts and fashion from them.

They owned Mathias Drapery in Rotorua, and in her spare time Grandma knitted bed-socks and bed-jackets with satin ribbons for the shop. Please try and imagine a little girl with black curls sitting up in bed decked out in powder-blue bed-socks and a pastel-pink bed-jacket, reading childhood horror storybooks where everyone gets tortured and eaten, usually in forests, while simultaneously sucking on Grandma's cough lollies. When our family visited Rotorua, their drapery shop was a realm of heavenliness for me with its shelves of fabric, curtains, household goods, sewing and knitting necessities, patterns, clothing and wooden stairs leading up to the huntin' fishin' shootin' department. My grandparents were both always beautifully dressed; I never saw them in sports or casual clothes. My father worked in department stores when he left school, and in his twenties was in the fabric department of David Jones in Sydney. Is it any wonder I am drawn to fashion?

Top: Embroidered whitework quilt by my mother; and below: my sister Keriann's exquisite embroidery and quilting.

Dancing dresses

The earliest dress I remember is my ballet tutu, and I mostly remember it because of the feel and smell of it and the ballet pumps. The pale pink satin bodice felt delicious against my five-year-old skin, and the sticky-outy skirt was magical because its stiffness appealed to my anal-compulsive side. I think Mum made it. I can barely even talk about those pink satin ballet pumps — the memory of the scent of the lacquer represents the joy and elation I found in dancing. For performances we wore make-up and had our hair curled. If I could have lived on a stage I would have. I just loved being up there in my pink tutu. As with all childhood concerts, our parents laughed till they cried at our haphazard attempts to be graceful and not fall into the lights. At one point, I refused to take the tutu off and wore it to bed. Mum didn't argue with me because, as she explained to everyone, I was highly strung.

The oldest dress I still own is my debutante gown made by my mother in 1966 when I was seventeen. It was not stored in my garden shed — I found it in my spare-room wardrobe. I have no idea how or why I still have it. I have moved houses, travelled constantly, lived for long periods in other countries, reincarnated myself as a hippy, a chic Parisian, an Indian princess and a Japanese pleat person and yet . . . and yet . . . somehow I still have that debutante dress. How can this be? Does anyone else in the entire world have their debutante dress?

I can only guess that when I left New Zealand for Canada in 1973, draped in a long flowery dress I had made, I probably stored some possessions with my parents. Subsequently, they moved to Australia to live and must have taken my treasures with them. That's not inconceivable because they took their whole lives with them from the day they got married, including the couch and chairs my grandparents gave them as a wedding present. They died in their nineties with that lounge set still in their living room. But they had six children and I was the eldest. Surely they didn't take *all* their children's things with them? Again, that's not inconceivable. I still have my grandmother's dressing-table hand-mirror, combs and brushes. I still have all the smocking

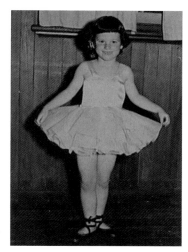

If Anna Pavlova can have a dessert named after her, surely this merits a *petit four* . . .

samples on Liberty fabric for the dresses I smocked for my nieces. I still have boxes full of ribbons, lace and buttons I bought in my twenties. I have shelves full of fabric I have bought all over the world, just waiting for me to decide what to make with them. Holding on to things is a way of holding on to the past, which I don't think is a bad thing because if you don't know where you came from then you can't know who you are — holding on to the past is deeply about identity.

The Catholic debutante ball's ostensible purpose was to present young maidens to society. We were introduced by the Archbishop of Auckland amid pomp and ceremony in the town hall. This wonderful old room of gilded angels, embossed rosettes, sweeping staircases and red carpet with fleur-de-lys welcomed 80 girls all in white and 80 boys all in black with bow ties. We snow-maidens were all led the length of the town hall on the arms of our adoring fathers. When I waltzed with my 6 foot 3 inch partner, my eyes reached his chest. He had a Catholic sense of rhythm — it felt great but it didn't work. Debutante balls started in the seventeenth century when high-born women were presented to court as their coming out — it was basically a marriage market. In New Zealand it was less élitist and more like the American prom.

Escorted by my
father, I show off the
debutante gown made
by my mother; and a few
decades later I can still
get into it. Just.

Mum made my debutante gown from white silk organza with tiny daisies embossed all over it. Right along the hem she attached handmade silk flowers with little diamantes in the middle. I begged her to lower the neckline, but she refused. She made a pale-pink satin petticoat to go under the dress, explaining that the colour matched my skin and that a white petticoat under a white transparent dress was obvious and vulgar. For that era, this concept was considered revolutionary and was much commented upon; needless to say, it has been in and out of fashion in the decades since. My long black hair was rolled into huge petals, silk flowers were stuck in it and the whole story was topped off with elbow-length kid gloves and silver satin pumps.

When I started nursing in 1968, to go dancing meant to go to a formal ball, which in turn meant glamorous gowns and special hairdos, often involving petals. If I made my own gowns, then I could decide the level of the neckline, so for my next ball I duly chose a pattern with a neckline that was not only low but also accented with a white ruffle. The gown had an empire-line and was made in navy-blue organza with tiny white dots. The standard make-up entailed black eyeliner, brown arcs of eye shadow, false eyelashes and pale pink lipstick. False eyelashes were only for significant balls, for less formal occasions three layers of mascara was the only icing this particular cake required. As I tottered off to the ball, the father said, 'Careful you don't lose your balance, girl.' The mother said, 'You look much better without make-up, Peta.' What did she know? I wouldn't have even considered leaving my bedroom without make-up. Another ball gown, which I wore with elbow-length gloves, had a plunging neckline and a crinoline skirt that wouldn't have looked out of place in *Beauty and the Beast*.

The next oldest dress I own is the one I bought for some nursing ball or other when I was 18 — a Mary Quant-style chocolate-brown silk organza with pin tucks made by Jennifer Dean. The label says PSSW. (I don't know what that means, but I'm sure it doesn't refer to a Psycho-Social Support Worker, which is one suggestion thrown up by the internet, and coincidentally was something I would be doing a few

Belle of the Ball (the Beasts were staying out of view).

years later.) The dress is so tiny you wonder how even a child could get into it. This dress was much more funky than the formal ball gowns worn by my contemporaries, and I felt quite groovy in it.

Influential eras

One can't talk about the 70s without mentioning the exuberant haute-couture designer Vinka Lucas. I couldn't buy her clothes because they were so expensive, but any self-respecting bride or wealthy woman worth their diamonds was decked out in her flamboyant designs. She also started up a magazine called *New Zealand Bride*, just in case you couldn't stop in at her spectacular shop on Queen Street. New Zealand was so boring in those days that people didn't even know they were bored, so when red-haired Croatian Vinka, with her strong accent and European background, hit the scene, she seemed like a peacock from another world. She was put down by tall-poppy New Zealanders a lot, but she just kept going and didn't change her sumptuous, theatrical styles one single bit.

Vinka had trained in Zagreb and was a brilliant cutter — she could throw eye-wateringly expensive silk guipure lace on the table and just

Rocking the Mary Quant dress when I was 18, and it is still going today; here paired with Moroccan earrings.

cut it into a dress, never making a mistake. She was a hard worker and clever businesswoman, designing bridal dress patterns and stocking unique imported fabric so people could make their own if they wished. She had brides in hot pants, giant hairdos full of flowers, leopard-skin umbrellas, elaborate head-pieces, sparkling gowns studded with stones and embroidery, and ostentatious chandelier earrings. Vinka died in 2020, and there will never be another designer like her. Her funeral notice advised mourners: 'Vinka's standard of dress mandatory'. Her daughter Anita still runs the shop on Queen Street, now called Vinka Design.

The 70s has been described by some as 'the decade that taste forgot'. I was there and thought it was fabulous because we were liberated and released from all sorts of things: fear of pregnancy; control of the patriarchy; sexual repression; out-dated fashion and dressing rules. How could a kaftan and headband be wrong? And those denim flares with wedges and fringes hanging everywhere? Believe it or not, flares are coming back again — shut the front door! And jumpsuits — stop! And the shaggy Afghan coats.

In the 70s it didn't take long for me to become fully engaged in the hippy revolution, but most of my straight nursing friends weren't. How could this be? How could you live in society, be at a formative stage in your life and not be moved at all by what was going on all around you? How could you carry on with the old values and conservative mindset when you were being offered freedom, excitement and unprecedented supplies of alfalfa sprouts? How was it possible that a girl would continue wearing tasteful camel-coloured Angora twin-sets, pearls, pantyhose and hairspray when she could be decked out in macramé coats, antique white lace from the Salvation Army shop, sandals and beads?

My most criminal fashion behaviour involved the aforementioned wearing of exactly the same dresses as my sisters (technically my mother's fault), sporting hair-pieces that didn't exactly match my natural hair colour, appearing in false eyelashes at the beach, strutting in white stilettos, owning paisley trouser suits (also technically my mother's fault)

Black wedding dress, red veil, feathered collar and shoes that were more sword than stiletto (sadly, the red heels are hidden in shadow).

and wearing macramé hats (why didn't someone tell me I was wearing a plant holder?) — so obviously a lot of these occurred in the 70s. 'A little bad taste is like a nice splash of paprika,' fashion editor Diana Vreeland was reported as saying, bless her. 'We all need a splash of bad taste — it's hearty, it's healthy, it's physical. I think we could use more of it. No taste is what I'm against.'

In 1980 I went to Paris for three days and stayed for 10 years. Oddly enough, it was the men in my life who influenced the way I dressed in Paris during those 10 years, not the women. I will write later about my boyfriend, Alain the chef with the beautiful shoes, but my husband Alexy also influenced me. Alexy was very handsome in that Nureyev way: high cheekbones, classic square jaw, polite, gentle and charming. He was a model and carried himself like a model, as if he were showing an Yves Saint Laurent suit. Even when he was just wearing jeans and a shirt, the jeans fitted perfectly, the shirt collar was half up and he wore perfume. When we got married, he wore a vintage morning suit and a magnetic smile, and I wore a black dress an Italian designer friend had loaned me. This I paired with black patent leather stilettos with red heels and a blood-red veil.

If you've read my previous books, you'll know that after about a year in the City of Light, I was invited to a 'salon' at the gay German photographer Willy Maywald's atelier in Montparnasse. At that time Willy was in his late seventies, he had been hosting this salon every Tuesday night for the past 30 years and it was crawling with artists, writers, filmmakers, fashionistas, ancient Russian aristocrats and generally interesting people. He held court in navy-blue pants, striped shirt, always a tie, navy jacket, a cigarette holder in his long elegant fingers, legs languidly crossed. Mocking blue eyes smiled out of his handsome patrician face, his silver hair brushed straight back. He looked impeccable, and he made me want to look impeccable. We subsequently became friends, and I ended up living with him.

Willy was famous for his photographic portraits of painters and writers (in fact his atelier where he also lived was previously owned by Modigliani), but he was even more famous for his haute-couture fashion photography, being a master of composition, lighting and subtle elegance, even if the models were in the arcades, footpaths or stairways of Paris. He was the principal photographer for Dior, and also worked with Jacques Fath, Yves Saint Laurent and Piguet. Don't tell the nuns, but *Vogue* was my Bible from an early age, and here was someone whose photographs graced its covers and those of *Vanity Fair*. There was a huge retrospective of his work at the Fashion Institute of Technology in New York a few years before he died in 1985, and the year after he died there was a year-long exhibition at the Musée de la Mode et du Costume in Paris. Google his name and you'll get a taste of his fashion photography over much of the twentieth century: wasp-thin waists, statement hats, gloves, tiered gowns and suits tailored into every contour.

Willy's home was on one of the most famous streets in Paris: rue de la Grande-Chaumière. Just walking along this little street, you immediately felt artistic and bohemian. The Académie de la Grande Chaumière art school was founded by a painter in 1904 to teach both painting and sculpture, but not in the restrictive academic way of other institutions of the time. Picasso, Delacroix, Manet and Cézanne painted there. It was still operating when I was living in Paris. My mother, who was a painter, spent hours in their art supplies shop when she visited me. This was the atmosphere I soaked up and the world I mixed in, so I had to dress the part. But how when haute couture was well beyond my income? Well, Paris is a dream world for *dégriffés* shops (which sell marked-down designer wear with the label removed). My friends and I went mad in them. We had no money but always looked incongruously well dressed.

The fashion bywords of the 80s were irreverence, wit and unpredictability. I had been embodying all three since birth, and by trying to make me do the opposite, the nuns ensured I'd keep ignoring them for eternity. Not that I am stuck in the 80s, you understand, just those attributes are timeless. Trendsetters tend to latch on to elements that have been around all the time and give them new focus as if they, rather than the focus, were new.

As I wrote in my book *Can We Help It If We're Fabulous?*, fashion writers can always be counted on to use the word 'story' at least 37 times in an article, which is why I have made sure the word has ample mention in these pages. Through the 80s we had such 'new' things as the English Tea Dance Story, the Decadent Opulence Story, the Bolivian Mountain Girl Story and the Dominatrix Story. Having now been to Bolivia, I know that Bolivian mountain girls wear the same dress and no knickers every day of the year so they can go to the toilet free-range — they squat wherever and the voluminous peasant shirts gently balloon out, then they stand up, give a little shake and continue on their way. Saris can function in a similar fashion. There should be a Pragmatic Fashion Story.

What the 80s Story seemed to be about was strong silhouettes (the dreaded shoulder pads and big hair), revival psychedelic (as I keep saying, recycling is inbuilt in the story of fashion, even when it is bad first time around), leopard-print platforms (ditto), and hot pants (for the handful of women whose thighs do not belong in a Rubens painting or conversely on a statue by Giacometti). The hot pants were nearly as bad as the Spandex-wearing — such acts against decency were almost enough to send me back to the dictates of the nuns. They would be pleased to have known that the 80s taught me not to appear at the breakfast table in transparent clothing, no matter what flattering lies my lover told me the night before.

In the 90s the knowledgeable fashion victim dressed for attitude, dressed to kill, or at the very least maim. Sexiness was hard and violent, power dressing was out, thank God, and contempt dressing was in. We're talking trashy thongs, spiky stilettos, patent leather straps, black leather, sex-shop styling, bondage detailing and fetish-inspired boots. Believe it or not, real fur was replacing fake fur and the fashion writers were saying 'dominatrix' *again*. Are we so lacking in ideas that we have to recycle those stupid giant shoulder pads, feathers and fishnet stockings? At least heroin chic is now out. I mean, how much purple eyeshadow can you wear under your eyes anyhow, and do you have any idea how difficult it is to peroxide your hair and deliberately leave the roots black?

Let's fast forward through the 2000s hip-hop, casual chic, retro and fast fashion to the 2020s, where the days of rules (beyond mine, of course) are out the glass ceiling. It's all deconstruction and, quite frankly, jolly anarchy. It could end in tears. Having been reared by *Vogue* in dogged enslavement to an impossible ideal, how can I now be expected to adjust to the fashion industry kowtowing to the consumer? This is role reversal at its most confusing. Young Parisian girls wear ridiculously high platform shoes, and the boys, who seem to be getting taller, wear tight black jeans and black T-shirts. Even though they fall over at the idea of exercise and appear to eat lots (unless perhaps they do it with mirrors), they remain slim and sexy.

Favourite designers

One of the reasons I got my first television job in 1994 was because of the way I dressed: they liked that I stood out. But it wasn't until I started my television career and was dressed by New Zealand designers Zambesi, Trelise Cooper and Scotties that my head and heart were turned by

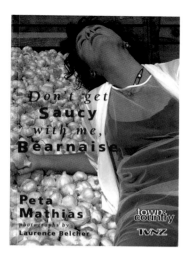

Don't get
Saucy
with me,
Béarnaise

Peta
Mathias
photographs by
Laurence Belcher

town&
country
TVNZ

If you're going to swoon
over garlic you may as well
do in it Zambesi.

off-the-rack designer clothes. They were just such a joy, exploding into my life as a side effect of having a clothing budget somebody else was paying for, or at least subsidising.

The iconic New Zealand Zambesi brand was started up in 1979 by Elisabeth and Neville Findlay. When I began presenting *Taste New Zealand* on TV One, Zambesi was the grungiest, most intellectual, most avant-garde designer shop around. If you wore Zambesi you were immediately telling the world you didn't vote conservative, you had a brain and you were sort of artistic but not trendy. Liz is not a look-at-*moi* sort of woman, preferring to remain in the background, and she loves black (probably because she's Greek, or maybe because she grew up in Dunedin), but she'll also throw mad splashes of colour in. She starts with fabric then invents something around that, the designs being layered, beautifully tailored and full of her memories of other clothes she has loved. She is the eldest of six children of a Greek/Ukrainian immigrant family and learned to sew and love fabric from her mother Zinovia. Zambesi clothes are still made in New Zealand.

Liz Findlay leads me to her sister Margarita Robertson, who has her own successful label, NOM*d, designing knitwear and woven clothes.

On my first-ever series of *Taste New Zealand*, I wore one of her jumpers on a fishing trawler. It was the most funky and unusual jumper I had ever seen and reminded me of Japanese designers like Miyake and Yamamoto — kind of utilitarian and conceptual. It was very heavy and squid-ink purple with extremely long sleeves that folded back on the arm, a roll neck and raised hemline at the front. I wanted to wear it with a netting skirt and Patrick Cox ankle boots I had from Zambesi, arguing that you should never lower your standards just because you are on a fishing trawler on Otago Harbour in high winds, but the crew wouldn't let me. I cooked a bloody fish stew on that boat in that jumper and I still have it.

Another New Zealand designer I worked a lot with in the television days was Marilyn Sainty. She was in partnership with Sonja Batt of Chez Bleu, which they had transformed into the first Scotties Boutique in 1979. Marilyn is another of these don't-look-at-*moi* designers, with a soft voice and gentle smile, an intelligent and sensitive woman who made very stylish, witty, elegant clothes, which were worn by thinking, artistic women. You'd splash out and buy a Scotties outfit if you had an important do to go to, like a funeral or a wedding. Marilyn's clothes were never sexy or in your face and were often made of hand-printed textiles. She and Sonja imported really amazing design labels to sell in their shop in Herne Bay; people like Dries Van Noten, Comme des Garçons and Ann Demeulemeester. Marilyn has retired and Sonja is still there, importing gorgeous garments and making sure the recycle department is full of unbelievable bargains. It is the always beautifully dressed Sonja who loaned me the Issey Miyake coat to go to New York in, Sonja who still has the eye of the master buyer she is and Sonja who manages to talk me into yet another recycled Marni. She's a truly stylish woman who knows how to show others what's what in the sartorial game.

Trelise Cooper burst onto the New Zealand scene in 1985 in a flurry of colour, romance, sequins, frills, tulle, flowers and opulence. She made women believe that they could look fabulous, fashionable and feminine

One of my Trelise dresses, exhibiting her exuberant love of pattern, colour and flounces.

no matter what their size; at that time she was a voluptuous woman herself so she totally understood that everyone wanted to have fun and be riotous, not just tall, thin girls who looked good in yesterday's flour sack anyway. Her belief in the transformative power of dressing well and looking good has never wavered. The Trelise Cooper clothes I wore on my television show were absolutely outrageous — she got women wearing flowers in their hair, striped bell-bottoms, sexy sundresses cut on the cross, silk skirts with rows and rows of frills underneath. She made everyone feel as if they were in a beautiful garden full of Rococo paintings. Her label is incredibly productive, sells all over Australasia, and my garden shed is full of her stuff. I may not wear it so much (although I did wear one of her iconic silk party skirts to my seventieth birthday party, and a gorgeous orange beaded evening dress on stage in my one-woman show to do a cooking demo), but I can't bear to sell it either. Trelise designs are living proof that you may change your taste over the years but if you really want to stun the room you will go Trelise.

Among the smorgasbord of overseas designers, I have had obsessions with a select number over the years: Jean Paul Gaultier, Prada, Péro, to name a few. One of my longest passions has been with the glorious

Look carefully at the brown and white pattern of my Marni dress and you'll see a flock of flying birds.

Italian brand Marni, founded in Milan in 1994, the year I started in television. Marilyn Sainty, with her impeccable taste, introduced me to Marni and also to Issey Miyake's brand Pleats Please. When the salespeople at Scotties see me coming, they take me straight to the Marni recycled clothes, bypassing everything else. I imagine there are a few rich ladies in New Zealand who are supplying me with these beautiful dresses.

The unassuming Consuelo Castiglioni was the avant-garde, creative force behind the Marni label of husband Gianni and six other family members. Italian designs are typically slick, sexy and sparkly, but not so Marni, which has a loose, round-cornered, feminine and unrestrictive style. The clothes feel soft and are so comfortable to wear, make you look beautiful without you feeling exposed or over-sexy, and their unusual painted prints and appealing colours can be mixed and matched with previous seasons' pieces. With handmade, almost crafty details, this is an eclectic label. I regularly wear Marni dresses I bought 20 years ago and they still look gorgeous. I buy them all over the world in sales and second-hand designer shops. Walking into a Marni shop is like walking into a colourful art gallery — the clothes dare you to start a subversive revolution in your life, to go mad and wear something intelligent and eccentric.

I particularly love my 2010 Marni brown and white dress and wore it in many photoshoots, notably in *The Great New Zealand Cookbook*. In this book, I wrote that I discovered when young that, if you can cook, people will like you. Being able to cook means you can share something of yourself that doesn't cost much, and the magic of recipes is that they connect you to the past. I contributed my father's roly-poly pudding — an old-fashioned pudding made with suet rather than butter — and also duck *confit* and *pommes de terre sarladaises* (potatoes in duck fat). I love fat and never cut the fat off meat for fear of losing my curves.

When I'm in Paris, London or any big city that has a Marni shop, I always do a pilgrimage just so I can stand in the chic beauty that is a Marni zone, and so I can drown in all the colours. If there's a sale on, the situation gets pretty dangerous. Consuelo sold Marni to Renzo Rosso in 2016 and now the creative director is Francesco Risso, who has changed things quite a lot, but I still like it.

Another international designer introduced to me by the Scotties women is Japanese Issey Miyake, particularly his Pleats Please range. Even though I very seldom wear black and would much rather have bought the extraordinary Pleats Please bat dress in another colour, I couldn't resist the shape and Miyake's sense of humour. I found it on sale in the Pleats Please shop in London in 1998. It looks great with my orange Pleats Please pants and lots of wide colourful acrylic bangles, Iris Apfel style. I have worn this dress all over the world, and even men come up to me in restaurants and say 'Pleats Please'. This is pretty impressive considering a man wouldn't notice if you were wearing a shroud with fairy lights.

I wore lots of Pleats Please dresses on my television shows over 12 years, to the point where viewers wrote in asking if I could wear something else. These stretchy pleated clothes thrilled me — how could anyone be so clever as to invent clothes that were wearable art? I mean really practical and wearable and indestructible but also soaringly

The Pleats Please
bat dress may
be black but it's
stunning and a hoot.

beautiful, exciting and architectural. You don't have to look after them, don't have to iron them, you can play with them, twisting them in and out of each other to create different looks, and they magically suit every occasion from coffee at your local to front row at the ballet. These clothes are the perfect solution for a woman who works and travels — they are light, you roll them up, tie them in loose knots and pack them flat. They arrive wrinkle free and perfect. Also they are easy to store at home because you lie them flat or knotted in drawers rather than on hangers. Most of these clothes I bought in the 1990s, in different countries in the world, usually on sale, and they still look the same and I am still wearing them.

It's a beautiful story. Miyake experimented a lot before he invented his own fabric to get the Pleats Please look — the first collection came out in 1993. He developed a quality of polyester that could be pleated with a special heat system, thus remaining permanent (more or less — the pleats do loosen a little over time if the garment is worn a lot). This tight pleating and silkiness of the fabric allows the clothes to flatteringly cling to the body while simultaneously giving you lots of room and air between you and the garment. Pleats Please are not just pleated; they are folded, cut, fringed, snipped into lace, patterned and sculptured, which is why the fabric has to be synthetic. As with smocking, to make a garment, the original size has to be three times bigger than the finished pleated product — this requires a lot of cloth. So you make the garment three times bigger, then you print your design on it, then you pleat it so the design shrinks to a normal size.

Cut on the bias, this Trelise Cooper
sundress would enhance most figures.

Dressing like a Parisian . . . or an Italian

I have always loved Paris, but not everyone does. When I was interviewed by French journalists about my book on France, called *Salut!*, they thought I was suffering from some sort of altered state. '*Mais les Parisiens sont stressés, moches, agressifs. Ils n'aiment personne, surtout pas les Parisiens.*' (Parisians are stressed out, ugly and aggressive. They don't like anyone. They don't even like each other.) 'No, no,' I said, 'you French are so negative. You all have good educations, you live in a country so beautiful that you have the most tourists in the world [89.4 million per year at latest count] in spite of your efforts to humiliate and alienate them, and you have arguably the best cuisine and fashion. But still you complain. Still you can't allow yourselves to be happy.' 'It's true,' they replied, 'we're very insular and it's revealing to hear how foreigners see us and how kind you are to us. We don't deserve it.'

Learning how to dress like a Parisian is quite an art. To give me some insights, I bought a book called *Dress Like a Parisian* by Aloïs Guinut. Aloïs studied fashion at the Institut Français de la Mode and runs her own personal shopping and style-coaching service. She doesn't believe in restrictive fashion rules and thinks you should use fashion to enhance your personality rather than shaping your personality to fashion. Fashions fade, style is eternal, as Yves Saint Laurent always said. As part of my research for this book, I decided to book one of her fashion tours in Paris.

Of course, you think you know Paris like the back of your hand when you've lived there for 10 years, visited it every year since and written two books about it. *Pas forcément.* For instance, I reckoned I knew the places to haunt, such as around rue des Petits-Champs, which is a wonder world of ancient passages, notably the very chic arcade called Galerie Vivienne, or rue des Saints-Pères in Saint-Germain. In this area I used to check out Sonia Rykiel's shop, which had clothes that would look good on tiny, thin, pale, red-headed women just like her: slick Yamamoto; garish Versace; Anne Fontaine, selling only white blouses and shirts for women — crisp, pristine and cleanly Catholic;

and classic Yves Saint Laurent. Having clocked everything, I would then treat myself to a coffee at the famous Deux Magots café (operating since 1885) on the Boulevard Saint-Germain. The waiters are straight out of central casting with their black vests, floor-length starchy white aprons and handle-bar moustaches. It was originally a literary café, but now it's mostly the fashion and political world who grace the velvet banquettes.

However, I had new tricks to learn. When I met up with Aloïs outside the Saint-Paul Métro in the Marais, she looked exactly as I expected her to: slim, young, pretty and chic in a casual way. Dark hair and dark eyes. It was the end of winter but exceptionally cold. I was on my way from Vietnam to my home in the South of France, I had no warm clothes and was suffering from both gastroenteritis and a bad cold. I was dressed in the most incredible combination of layers and layers of summer clothes and would not have got out of bed had it not been for my date with the lovely Aloïs. I certainly didn't embody her seven rules of style:

- effortlessness — you should always look like you just woke up like this

- elegance — otherwise known as everyday visual politeness (this is my favourite)

- understatement — have attitude but be discreet

- sensuality — reveal what's already there rather than add to it or hide it

- fun — be witty, be playful, be light-hearted

- individuality — singularity lies in the details

- breaking the rules — to be Parisian means to be free-spirited.

My vintage
1980's Roger L.
Paris souvenir
scarf under a
spewing swan.

I am happy to say I managed to break all the rules that day with my weird look. Oddly enough when we walked into the first shop of the tour, the owner said, 'Oh my God, she looks fabulous. Who is this woman?' I looked at Aloïs, who said, 'He's talking about you.' This may or may not be why I bought two pairs of shoes in his shop.

Here's another piece of advice from Aloïs. So you're wearing jeans and an oversize sweater and are feeling frumpy. Solution? Crop the jeans, roll the sleeves up, throw on some sexy heels and a chunky bracelet and boom, you've got that effortless French-girl style. Personally, I think there is no point in doing one of Aloïs's fashion tours if you're not open and really wanting to learn about Parisian style. I read the most ridiculous story written by a New Zealand journalist where she obviously had no idea and was not interested in getting an idea. We visited all sorts of places: workshops, private sales, up-market shops in the Marais, lots of shoe shops and perfume shops where you get to go out the back. It was so interesting. We walked, got taxis, jumped on the Métro, had tea breaks with couture pastries and met her friends. Aloïs has now put out a second book called *Why French Women Wear Vintage*.

Another good place in France to pillage for ideas is Nice. The

Famous for creating sand-blasted jeans, Italian designer Roberto Cavalli also favoured bold prints, as you can see in this silk jersey dress.

Niçoise can always be relied upon to access their inner-Italian fashion style (Nice used to be Italian, and they have never forgotten it, either in fashion or in gastronomy). Think the bombshell look with form-fitting clothes, plunging necklines, over-tanned, gold jewellery, bottle blonde, exotic prints, sand-blasted jeans. But just because they like sexy doesn't mean they like totally-in-your-face; in fact Italian women like to show off their curves, not their skin and not their least attractive assets — whatever's good they highlight, whatever's not so good they hide. A little bit of glitz goes a long way. I found an elegant, sexy silk jersey dress in the Roberto Cavalli shop in Nice. I wore it with high-heeled sandals to the *corrida* in Arles with my friend Gina, who was dressed appropriately in jeans, boots and Souleiado shirt. I think next time I might wear jeans — the bulls didn't even notice my look.

Italian women are unbelievable. If you don't agree with me, go to an Italian wedding. Stand outside any Italian church when the wedding party is emerging. I went to a four-day one in Puglia and everybody, including men and children, looked like they were on their way to the Oscars, in different outfits, not only every day, but several per day. I wore an embroidered lime-green dress with Jimmy Choo shoes and chandelier earrings and felt under-dressed. Women of all sizes and shapes were dressed in bombshell, chic, refined, sexy, dramatic clothes. If there's one thing that will make an outfit, it's fabulous shoes, and in Italy that means high and pointed. Old grannies at this wedding were wearing stilettos — they removed them to get up the church steps, then put them on again. They also had perfect make-up and hairdos. In a country where the police uniforms are designed by Armani, it's not surprising the men know how to dress too.

Stalking the famous

Let's talk about aspirational dressing. You might not feel comfortable in, say, a tight sexy dress because you've told yourself all these limiting things about your body. Sometimes if you do something really scary, it works just because it gives you self-confidence. I'm thinking of the Australian actress/producer Rebel Wilson who is very voluptuous and positively rocks those curves in clinging clothes, to the point where she has her own fashion label: Rebel Wilson x Angels. Even though she's now lost a whopping 28 kilos (at time of writing), she's still curvy in all the right places. Alternatively, you might like to emulate the pared-down elegant look of Audrey Hepburn. Elegance is a good aspiration because it is an attitude and, as Aloïs says, a form of visual politeness. If you go out looking like a dog's dinner, you're just being rude to other people.

There is, though, evidence that rudeness, or even aggression, stops us from expressing ourselves through our clothes. In a 2020 report published in *Social Psychological and Personality Science*, J.A. Krems, A.M. Rankin and S.B. Northover concluded that women expect more aggression from their fellow sisters if they are dressed 'revealingly versus modestly, especially if targets are attractive' and further that 'women create outfits baring less skin, select more modest clothing, and intend to dress less revealingly to encounter other women' to avoid this aggression. I'm not suggesting you bare all, but I'm saying we should all be able to wear what we like (within reason). I'm also saying to the

CHAPTER 4

New York

While I was writing this book, my friend Johanna's daughter, who is a fashion buyer in New York, visited for Christmas and we got talking. Six-foot, 27-year-old beauty Greta, who did her New Zealand degree in fashion and design, was decked out in a dramatic transparent eyelet-lace Lisa Marie Fernandez dress with a double ruffle on one shoulder neckline. Boom-boom. This is Greta all over. When she was a teenager, she was always buying fabulous second-hand clothes online. One day she turned up to my place in a see-through top and vintage silk shorts, which were basically underwear. Just before I opened my mouth to make a comment, I remembered I'd done the same thing at that age. My mother's spies would call her to report they had seen me walking around in my underwear (vintage lace from the Salvation Army shop *actually*).

So Greta grew up and the one thing on her mind was getting into the fashion business in New York. She was telling me about her life there, and it entered my head that here in front of me was a private guide to the fashion scene in that great city, a situation I had to take advantage of to add another dimension to this book. So we hatched a plan, I took some notes, Sonja Batt from Scotties Boutique loaned me her Issey Miyake fake fur coat, and the next thing I knew I was in the very chic and funky Soho Grand Hotel in the most happening part of Manhattan, waiting for Greta to change my sartorial life.

It was one degree Celsius and we hit the street. I was *so* ready in my new winter shoes, cashmere socks and the same clothes I had worn on the plane. Why waste time with boring showers and ostentatious outfit changes? New York restaurants are good but not desperately imaginative, and Greta soon got impatient with my high standards. We walked to dinner at the last word in sophistication, the Mercer Kitchen. Everything in New York is industrial modern (all copied from my house in Uzès, just saying). The food was fine but nothing to lie down and die for (unlike at my house in Uzès, just saying).

Following dinner we visited Café Select for a drink, and I made my first social faux-pas in New York by ordering a cosmopolitan, just like Carrie in *Sex and the City*. The barman looked at me incredulously as if to say, *What time warp are you from, honey?* I asked Greta what the attitude was for, and she told me coolly that no one drinks cosmos — it's very dated. From that moment on I took to martinis. Time went by and we ended up at the Neo-Baroque Paul's Baby Grand cocktail lounge in Tribeca. This is a small private club known for its tough door policy, but Greta knows *everyone*, so we found ourselves rubbing shoulders with the great and the beautiful, the famous and the funny, *not* drinking cosmos. Cell phones are prohibited, the décor is floral, drinks are served on silver trays and there's a formally dressed doorman for the toilets (presumably to control substances going up people's noses). I loved it. I wanted to move in.

In my quest for fashion and clothing inspiration, I actually attended an organised *Sex and the City* tour, driving around visiting landmarks from the show. It was okay, but I expected it to be a lot more about fashion and shopping. The highlight was visiting the sex shop Samantha got her supplies from, where I found myself with a coffee in one hand and a dildo in the other. Couldn't think what to do next, so took a selfie.

Being in New York felt wonderful. It was not as I had remembered from my previous visit in 1976 — it was a dump then. I think I stayed in the Bowery, which was really rough; my friend had bars on her windows and skylights. This New York was clean, rich, ordered and seemed in control of itself. I thought the people would be aggressive and pushy, but they were the opposite: friendly, polite and very helpful. I happened to be there for Valentine's Day, and we had made absolutely no plans for dinner. All the good restaurants in this town are pumping, nobody appears to eat at home and you have to reserve for everything. I often saw queues down the street outside restaurants and clubs. But because Greta knows everyone, we squeezed ourselves into a new Greek restaurant, Lola Taverna in SoHo, where we ate very average food accompanied by extremely loud music and screaming patrons. The owner did take pity on us and gifted a fabulous bottle of Gevrey-Chambertin.

In winter, New Yorkers tend towards casual, layered clothes that they dress up with big scarves and fancy boots. They also wear luxury active and sportswear, which is so beautiful it doesn't look like they have come straight from the gym. This is in spite of the fact that Karl Lagerfeld once declared that 'sweatpants are a sign of defeat'. What was seriously trending both in New York and Europe when I was last there was what has been called athleisure, where you pair leisure wear with regular clothes, for example, a tailored hoodie under a pants suit with trainers, a sweatshirt under a skirt suit with sandshoes, nice track pants with a cashmere jumper or jacket, a cross-over yoga cardigan with a big flowery skirt and chunky boots. A case in point was Greta's friend Robyn Berkley, who dined with us dressed from head to toe in her company's fashionable activewear.

I paid her a visit the next day in her business headquarters in a gorgeous white SoHo loft. Robyn set up the clothing business Live The Process with her New Zealand business partner Jared Sharma (of the New Zealand Ruby fashion family) in 2013. She was previously

If there's a Marni shop, then of course I have to enter it.

in public relations and taught yoga, and he was a stressed-out (now chilled-out) banker. I touched every single beautiful pant, cardigan, legging, top and bra, and bought a few so I could be desperately chic at my next yoga class. Already I am the only person in my class wearing lipstick and perfume, and I feel standards must be maintained. Robyn told me she wanted to create active leisurewear that could be taken from the studio to the street. She does a lot of yoga and couldn't find exercise clothing she liked. The people who buy her clothes like understated luxury, which is actually a description of how New Yorkers dress in general. According to Robyn, her buyers are people who don't like branding, but like quality, like ethically produced and like US-made.

No more is there a garment district in New York, so most clothing designed in New York is made in Los Angeles. Live The Process know the names of all the people who make their clothes — nothing is outsourced. The yarns are 40 per cent sustainable at this point, and they are working on doing better. The cotton is approved by the Better Cotton Initiative and the polyester is recycled. All the garments are two-way stretch, which holds you in, lasts for years and years and doesn't fade. They can be mixed and matched with previous seasons' designs. For Robyn, it's all about the workout within, so it's important to feel good on the outside as well as the inside, and part of feeling good on the outside is looking good. You may not be feeling great at all inside, so feeling beautiful clothes against your skin makes it easier to face the world.

On the way back from Robyn's, and just before closing time, Greta and I popped into the minimalist Issey Miyake Pleats Please shop on Prince Street in SoHo. I was just looking, okay? Greta was dressed to the nines: every day she turned up in another stylish outfit. I was hidden under the fake-fur Issey Miyake coat. The previous year I had seen a video of a gorgeous parade of the 2020 spring/summer Pleats Please collection, with dancers twirling in big swirly dresses. I didn't see these dresses anywhere in the shop, so asked about them. The reluctant sales assistants, whose lack of interest in selling me anything

This Matisse-like print on an Issey Miyake dress was crying out to be displayed in an art gallery . . . so I duly obliged.

was almost soporific, finally admitted that there might be one in the basement. They brought up a coral and white dress with a large Matisse-like print in very light pleated fabric. I put it on, started swaying as it floated around me and immediately turned into a work of art. The rest is history. When we visited the Whitney Museum of American Art the next day, I wore it with black tights and black shoes. A New York fashionista, good at juxtapositions, might have chosen thick white trainers or stroppy studded Doc Martens. As luck and art would have it, there was a huge shocking pink wall at the Whitney where Greta took pics of me flattened against it like a swatted orange fly. Maybe *un peu trop*, but the museum visitors appreciated it.

New Yorkers are in love with Italian and French cuisine and hamburgers in equal measure. If they go to an up-market restaurant they will order an up-market burger. I had lunch at the famous Gramercy Tavern with my lovely friend Rose Levy Beranbaum, America's sweetheart of cake and bread recipe writing, and she ordered a hamburger, her favourite dish at this restaurant. Another day, Greta and I had lunch at Sant Ambroeus, a Milanese restaurant in the West Village, and, as with any other popular restaurant, already at midday

necessarily reliant on selling horrendously expensive outfits to keep afloat. Clever. Turns out they were the precursors to what we now all have to do because of the economic destruction as a result of Covid — the pivot. Who knew that pivoting (shifting to a new strategy) would become the catch word of 2020?

Incidentally, guess whose business exploded and did very well as a result of lockdown? Why, Robyn and Jared's Live The Process of course, because everyone wanted gorgeous stretchy at-home clothes.

Needless to say, September 2020's New York Fashion Week was digital, several big names passed, and mid-range designers did what they wanted when they wanted without fitting into the Fashion Week schedule. Suddenly, because of Covid, everyone was pivoting, and what seemed like unbreakable old ways of doing things were in fact completely replaceable. Fashion will survive — it always does. The future seems to be the fringe: new ideas and innovative design will come from the small businesses because now they can afford it and they are really good at social media. And forget Berlin, forget Paris; think Africa, think India.

My Banjanan dress sits happily with my pink-edged settee, but even more happily with my friend Mimi and a glass of champagne.

CHAPTER 5

Colour

Spring is tender, green young corn and pink apple blossoms.
Autumn is the contrast of the yellow leaves against violet tones.
Winter is the snow with black silhouettes.
But now, if summer is the opposition of blues against an element
of orange, in the gold bronze of the corn, one could paint a picture
which expressed the mood of the seasons in each of the contrasts of
the complementary colours . . .
—Vincent Van Gogh, from a letter to his brother Theo

As you were probably taught at school but have long since forgotten, there are three primary colours — yellow, red and blue; three secondary colours, created by mixing two primary colours — orange, green and violet; and six tertiary colours, created by mixing a primary colour with a secondary one — red-orange, yellow-orange, yellow-green, blue-green, blue-violet and red-violet.

Interestingly, there is actually no such thing as colour — colour is just our perception of light waves hitting an object. For example, a blue dress will absorb all the wavelengths of light *except* the blue ones, and *therefore* blue is what we see. Some scientists think that the first humans could hardly see colour at all — they saw it dully and perhaps just saw black, red and white, and maybe, as time went on, green and yellow. Homer described the sea as 'wine-dark' — he may have been colour blind, or wine might have been blue in those days, or perhaps he was just being lyrical about the inkiness of the sea's depths. It may be, though, that the way we view colours has changed.

'Colours are the mother tongue of the subconscious' according to Carl Jung. We are all attracted to certain colours, which may have to do with profound feelings, held ideas, childhood memories, spiritual connections, cultural traditions or emotional states. For instance, I would be more likely to fly to the moon than wear a dull olive-green, but lots of people love this colour. For me it is muddy, unflattering and depressing, but for someone else it is soothing and earthy. If you are feeling down, wear colour in a humorous way and it will lighten your mood: if you're surrounded by grey, you will feel grey and without solace. You have to experiment when you wear colour — sometimes unexpected combinations will go well together and sometimes clashing patterns will sing. Gone are the days when you were forbidden to wear pink and red together or blue with green.

We are particularly influenced by the colours our parents and grandparents wore. When my artist mother painted the kitchen door shocking pink and wore a burnt orange suit, I knew I was onto something; I knew it was okay to be brave with colour and that colour would make me happy. We are lucky these days because we can get and wear any colour we like, but before the nineteenth century accessible and affordable colours were mostly limited to earth pigments and plant dyes, so you got terracotta, dull-red, green-grey and black mostly. Really unadulterated sharp colours were very expensive to produce, until the Victorians discovered and developed a whole range of new chemical dyes from such delightful sources as arsenic and distilled tar.

Getting your colours done or 'read' by a professional was very popular in the 1980s, and even today there are colour consultants around and an array of confusing blogs on reading your colour. I never succumbed to it because I didn't feel I needed to pay someone to tell me what my eyes could see and heart could feel. If I put on a muddy-green dress, I want to slit my throat. If I put on a pomegranate-red dress I feel like dancing. It's bad enough having a big nose and 30,000 freckles — why would I compound that by adding ugly colours? Also, I deeply believe that anyone can wear any colour; it's just a matter of shade.

I spoke to a colour specialist who said it is not only colour and shade you need to consider but also contrast. So how do you find out which shades suit and flatter your skin, eyes and hair? And what happens if your hair is not your natural colour? Apparently that doesn't matter as much as your skin, which doesn't change throughout your life, though your pigmentation will, so, correspondingly, the colour tones that suit you will change. The word is we are all either warm toned or cool. An easy way to find out your natural skin tone is to look at the underside of your arm; if the skin is pale and sort of pink and your veins look blue, you are probably a cool tone; if the skin is flaxen and creamy verging on peach with green veins, you are a warm tone. Brown- and black-skinned sisters are all versions of a warm tone in principle, but *not necessarily,* e.g., Halle Berry is brown but her skin tone is considered to be cool, and Nicole Kidman is white but her skin tone is considered warm. See how confusing it is? Maybe I do need to employ a consultant after all . . . or maybe that is the cunning plan.

Another test is to hold gold and silver fabric or jewellery up to your face. Gold lights your skin up and makes your eyes pop if you are warm skinned, and silver works if you are cool. I am pale, but with all the freckles and fake red hair I am convinced I look fabulous in gold. Cool skins look good in all shades of blue, and warm skins look good in yellow-based shades. Colour consultants used to love dividing us into weather systems, so if you are cool then you are summer or winter; if you are warm then you are spring or autumn. Some people are several, while

I deeply believe that anyone can wear any colour; it's just a matter of shade.

others are all seasons and prepared to weather the storm, but there will usually be one season that dominates. Basically, if you're summer, you're holding hands with mid-shades of blues, purples, turquoises and pinks. Winter skins suit stronger, crisper versions of summer colours. Spring people should go for light and bright shades, like pale yellow, dusky pink, true blue and, on a good day, a bit of orange. If you're autumn, stick to autumnal colours, like dark reds, chocolate, gold and burnt orange — basically a chocolate and fruit bar — and a touch of green. Madame de Pompadour (mistress of King Louis XV) was probably a summer as she made light green, powder-blue, pale yellow and pink the height of fashion in eighteenth-century France.

But wait. Just when you thought it was safe to leave the house, the game has become even more sophisticated. These days seasons are out and colourways are in — 18 of them. The Absolute Colour System was developed by Imogen Lamport and allows for a lot more subtleties, including neutrals. Apparently the eyes have it — you either wear a colour that suits your eyes (which enhances them) or you do a colour that is a complete contrast (which intensifies them). This is also where contrast in general comes in, so if you know how to wear contrasting colours together that flatter your skin, hair and eyes then you are really onto something. It's like a sliding scale and much more nuanced than the season system.

However, it's others who see you and find your colour schemes pleasing or not, so if you don't care what others think then you can wear shocking pink, iridescent red, yellow gold and ugly khaki all together all on the same day and sing 'I Did It *My* Way' all the way to the ball. Think of British designer Zandra Rhodes who, at 80, wears rainbow colours and even has bright pink hair. She says colour just makes her happy. She also says work makes her happy, which is why she's still doing it and why she is in such good shape.

Okay, so you're alone in the shop with terrible lighting, a tiny trying-on room and a shop assistant who is a pathological liar — how are you supposed to make a good decision? Use the pop technique. Lock the shop assistant out and do the blink test. Hold the colour up to your face and blink. If the first thing you see upon opening your eyes is the colour, then it's probably not for you; if the first thing you see is your face, then that is your colour and you will be wearing it rather than it wearing you. It's all very fabulous and dramatic to look like an exploded nuclear reactor, but even I think that lime green doesn't do most people any favours. There are four colours that colour specialists consider 'universal' i.e. to be able to be worn by all: melon, cobalt blue, off-white, and aqua blue.

And then there's personality. We all have our favourite colours that we love because our personality is attracted to them. These colours don't necessarily look good on us. Another technique to find out which colours look good on you is to listen to the compliments you get from other people. If someone says 'OMG, what a fantastic colour' that means the colour is wearing you. If someone says 'OMG, you look fabulous in that colour' that means you are wearing the colour and owning it. I'm not saying you should immediately join a modelling agency, but you should stand up straight, smile and say THANK YOU.

Red

Research has found that men find women more attractive when they wear red. In an article to test this, published in the *Journal of Hospitality & Tourism Research*, N. Guéguen and C. Jacob detail how they put waitresses in France in different coloured T-shirts. They found that colour made no difference at all to what female customers paid in tips. However, when the waitresses wore red, male customers paid them up to 26 per cent more than when they wore any of the other colours (black, white, blue, green or yellow). So, essentially red is a man magnet.

In my book *Beat Till Stiff*, I wrote about why I love red so much. I said that 'Red is the colour of life — if you are afraid of red, you are afraid of life.' It is the colour of both danger and sexual arousal. When you put red lipstick on, it just pulls your face together in one second. Strictly speaking, you shouldn't even put the rubbish out without lipstick on,

blood stains. Whatever the reason, it was pretty daft if you wanted to dodge bullets. Women, meanwhile, tended to wear blue. But then the gender associations reversed to what is more familiar today, with red or pink being feminine colours and blue being masculine. The lesson to take from this is not to let tradition dictate what colour you can wear. An earlier tradition would have probably said the opposite, and such dictates can only too easily become a straitjacket. Babies, for instance, look great in bright colours (why should they be doomed to pastels?), and you should likewise defy any colour injunction.

The kermes dye from scale insects reminds me of when I was in Bolivia and Mexico. There the peasants produce wool from alpaca, llama and vincuña and dye it with bright colours discovered by the Incas and Aztecs. These include crimson that comes from the dried and ground bodies of the tiny, white insect, cochineal. As with the kermes, it is the carminic acid the insects produce to scare off predators that produces the red pigment. The cochineal insect lives on prickly pear cactus plants, making them appear as if covered in white powder. I have picked off a bug and squeezed it hard, to see the deep red liquid ooze out: it is the acid, not blood. The locals manually infest the cacti with pregnant bugs, leave them for five months, then blow the creatures into buckets using compressors, before re-infesting the plants three months later. It's like watching people pick grapes in a vineyard. It takes 70,000 bugs to make 500g of dye. Unlike most other natural dyes, cochineal doesn't fade when fixed with a metal salt and is completely safe — it is even used as a food dye, proving once again that fashion and food go hand in hand. The invention of synthetic dyes pushed natural dyes like cochineal aside, but with recent fears about carcinogens in colourings, cochineal production is popular again.

My red Trelise Cooper sundress was bought for *Taste New Zealand*. Aside from being red it has polka dots, which are guaranteed to bring happiness — there's nothing like a dot to tell the world you are both playful and hardworking. There *is* a chance you will look like Minnie Mouse but you just have to go with it when sporting the polka. But it wasn't always like that. In medieval times, you would never have worn fabric with dots because they were reminiscent of disease; for example, the spots of plague, measles and small pox. In those days, it was impossible to weave fabric with evenly spaced dots (later machines would do that), so irregular dots looked like blood spluttered onto a handkerchief, as in tuberculosis. However, forget that, the origin of the word '*polka*' is Polish and refers to something small and feminine. Now there's a nostalgic thing to wearing dots — nostalgia for the 50s and 60s film stars like Marilyn Monroe in her polka-dot bikini and Liz Taylor in her powder-blue polka-dot dress, not to mention the Brian Hyland song 'Itsy Bitsy Teeny Weeny Yellow Polka-Dot Bikini', which brings me to the next colour.

Yellow

The magic of the sun transmutes the palm trees into gold, the water seems full of diamonds and men become kings from the east.
—Pierre Auguste Renoir

Although you could produce a lovely yellow from ochre and raw sienna, in its genuine form, yellow was unseen outside nature before the creation of chrome yellow in 1797. It became popular in the first part of the nineteenth century. Nothing matched the brilliance of the pigment made from the mineral chromium, which was quite orangey in its original form and also led to a rather bright yellow. Think of Monet's dining-room at Giverny in France — saturated in gorgeous sunny luminous-yellow happiness. And it wasn't just one yellow — it was many subtle shades, which is actually a great way to wear yellow in your clothes — a mix of sour cream, ochre, gold, butter, primrose. And what about the sumptuous pale yellow in Vermeer's *The Girl With a Pearl Earring*? Classically blue also goes really well with its complementary colours yellow and orange, especially in summer. Less classically, yellow and pink are very cute together, bringing a feeling of well-being and intensity (think of a gerbera with its yellow head and pink petals).

The navy-blue on top of this Péro dress lets me get away with wearing mustard.

A yellow dress is a symbol of hope and optimism — it sort of holds things together — and it's very fashionable at the moment, like uplifting buttercups in the dull landscape of winter. In India the colour is symbolic of peace and knowledge. Just wearing yellow boosts your serotonin levels automatically. It wasn't always that way: in the past yellow has been seen as scary, a symbol of betrayal, provoking over-stimulation (seeing yellow can actually make your heart beat faster and raise your blood pressure). It was also seen as sinful, contaminated or a sign of transgression — in the nineteenth century naughty books were bound in yellow. If you look closely you will see some yellow-backed books in Van Gogh's paintings. It's most horrible incarnation was the yellow star Jews were forced to wear by the Nazis — the ultimate symbol of stigma. We need to counter all that negativity; if you can wear it, then evoke daffodils, buttercups, sunflowers, lemons, egg yolks, citrines, rubber ducks, autumn leaves and sunshine.

Quite often we fall in love with a colour because of the association of it with someone we admire. Most recently the colour's fortunes changed when in 2016 Beyoncé made her video for the song 'Hold Up' wearing a drop-dead yellow shoulder plissé gown designed by Peter Dundas, the creative director of Roberto Cavalli. Strangely enough, almost any skin tone can wear yellow — you just have to wear the right shade. Fair-skinned people look good in mustard, like my Péro dress (combined with blue). Olive-skinned fashionistas can wear both pastel and vibrant yellows, and of course black sisters can wear absolutely any colour they like, including every shade of yellow ever invented — think Michelle Obama in that Schiaparelli haute-couture gown designed by Daniel Roseberry in 2019 — kind of an acidic yellow with crystal beading all over it.

Blue

Blue is a very old dye colour, originally invented by the ancient Egyptians — they mixed ground limestone with sand and azurite (a deep-blue copper mineral). In the fourteenth and fifteenth centuries, painters who used blue had to be very rich. Synthetic blue was invented in 1826. It is an ascending colour (meaning ascending in the colour spectrum) but also ascending in the sense that it lifts the spirits. It has the reputation of being cold, so is not often used in bedroom colour schemes, but it is cooling and refreshing to wear, and there are many very beautiful blues like ultramarine (meaning beyond the sea), Prussian blue, violet, cornflower, sapphire, lavender, powder-blue, cobalt and its descendent cerulean (originally made from cobalt magnesium stannate) and the stunning dark indigo. Vibrant ultramarine or 'true blue' was originally made from lapis lazuli and worth more than gold. Discovered in Afghanistan 6000 years ago, it is often described as the most illustrious, perfect, beautiful blue of all.

Blue is luminous, expansive and airy like the sky and the sea. In clothing, blue and white are quite dazzling together, reflecting light brilliantly. Blue is particularly charming and refreshing with red, especially in a contrast of patterns. It is said that the sort of people who like wearing blue are mysterious, polite, shy souls with leanings towards reliability and kindness. They are not a person who wants to attract attention to themselves, like those vivacious red wearers. Wearing blue

Miyake's colours are so striking: just look at this blue top, paired here with pants by Pearl in Auckland.

Majorelle blue took its name from the French artist Jacques Majorelle, who painted the walls of his garden and studio in Marrakech in this intense shade; the gardens are now open to the public and his studio houses the Berber Museum, which includes traditional costumes and jewellery.

also symbolises purity — the Virgin Mary always wears it, my school uniform was blue, my teenage choir wore it, my nursing cape was a fetching royal blue and I have worn it on occasions ever since, despite not quite fitting the shy criteria.

The most impossibly exotic blue I have ever seen is Yves Saint Laurent's house in the Majorelle Gardens in Marrakech. Marjorelle blue is an intense, clear, fresh and explosive colour that makes you want to just go home and paint your house the same colour, in spite of the fact that every other house in your street is white. Another lovely Moroccan blue features in the northern town of Chefchaouen, where all the houses are painted pale blue (like the pale blue Brahmin houses in Jodhpur, India). Meanwhile the Tuareg 'blue people' in the Moroccan desert wear indigo.

Blue makes you think of the beach (think of cerulean blue), but guess what the most flattering blue is, hands down, for any skin tone? Navy-blue — otherwise known as marine because sailors wore it — looks good on absolutely everyone in any season, which is why all those uniforms you hate are made of it. If blue is so calming and harmonising, why do we sing the blues and why do we say we are feeling blue when we are sad? There are various explanations as to the origin of this association. One is that in the old days ships would paint a blue band on the boat and put up blue flags if a senior member of the crew died. A threatening atmosphere used to be described as a blue devil. Singing or playing in a minor key or including flattened thirds and sevenths is synonymous with sadness — witness traditional American blues singing.

In fashion, designers don't use blue that much and, when they do, it's to make a point, maybe a political one (conservatives wear blue), or maybe a psychological one — psychics believe blue helps you tap into the higher levels of the brain. Redheads and blondes look scintillating in blue. In the 1954 movie *To Catch a Thief*, Grace Kelly was dressed by costume designer Edith Head in a stunningly cool ice-blue floor-length chiffon gown. It was an interpretation of a 'new look' Dior dress — fitted to the waist, with shoestring straps, a gathered skirt and floaty scarf. Even the handbag was the same colour. The whole outfit said 'look at me but don't touch me, I'm wealthy, fashionable and in control of my universe'. Of course Cary Grant was powerless to resist.

The story of indigo is quite a dramatic one. This inky blue comes from many different plants, predominantly *Indigofera tinctoria* (known as true indigo) — a shrub with very pretty pink flowers, small leaves and drooping seedpods. It looks like a sweet pea because it belongs to the same family. As opposed to lapis lazuli blue, which was very expensive, indigo grew in many places all over the world and for centuries was the only affordable blue available.

I find indigo blue rather grand, moody and subversive in its quiet beauty, in spite of the fact that it became quite utilitarian and is the colour used for denim jeans. To show you the power of this colour, the global denim industry is worth a whopping US$90 million. Denim fabric, or serge as it was called, was invented in Nîmes (de Nîmes = denim), then in 1873 Levi Strauss took this hard-wearing material to America and denim jeans were invented, originally for farmers. I have a friend in the South of France who has a huge country kitchen painted entirely in dark indigo. In the extreme heat of that part of the world, it is surprisingly cooling and also keeps the flies away.

For a colour that was so predominant for centuries, you would think it would be relatively easy to make but, *au contraire*, it's quite a palaver. Most indigo is produced synthetically now, but I am interested in the handmade, plant-based, natural version. The leaves are picked, then fermented in an alkaline solution such as lye. Then a whole lot of workers jump into the huge vats and agitate the water enthusiastically with oars to aerate it, producing a foamy green/yellow sludge on the surface. This sludge is then scraped off and sun-dried in blocks. When the fabric is dipped in the indigo and lifted out to make contact with air it turns dark green then, magically, indigo blue. The fabric can be dipped over and over again to obtain a really glorious deep colour. This hand-dyeing process means that the results will vary, not just because of us erratic humans but also because things such as the plant source, the climatic conditions (temperature, humidity, etc.) and even the bacteria on the surfaces and in the air can all add to the unique shade created. The dye ages very well and is the most colourfast of all the natural dyes; it doesn't require a mordant (dye fixer).

Indigo gets its name from India but, although there are 62 indigo plant varieties in the country, research has shown that it didn't necessarily originate there. The oldest known piece of cloth dyed with indigo was found in Peru, while other countries, including Japan, China, Vietnam, Korea, Iran and West Africa, have also created indigo dye since early times. India, though, was to become the source of indigo dye for Europe.

Indigo was a big money-maker and hugely important in global trade, involving restricted trade routes, cheating with substitutes, slavery, mistreatment and bribery. Just like the story of spice, before Vasco da Gama opened up the passage to India everyone thought indigo was Arabian. When the sea route was opened, the prices came down drastically. The incidence of slavery and miserable forced labour was particularly ignoble in India. Europeans (mostly Portuguese) first sailed into Goa on the south-west coast of India in the sixteenth century and bought up loads of indigo, spices and silks, making a huge profit upon

Belted by a bright scarf, this Injiri dress is made from hand-woven cotton dyed with natural indigo.

their return. By the seventeenth century, Indian indigo had lots of production problems with impurities, etc., and growth slowed.

There was big production in Bengal in the 1850s, rife with bribery, rape, bad conditions, mistreatment and brutality on the part of the British (under the auspices of the justly maligned East India Company). In Bengal alone, 4 million people were employed in the industry. This eventually resulted in the 'blue mutiny' and the beginning not only of the fall of indigo manufacturing but of the British Empire itself. Also, someone was about to discover a synthetic indigo. In 1856 William Perkin figured out how to make aniline dyes (an organic chemical compound), and in 1886 Adolf von Baeyer made indigo from coal tar, which is how chemical indigo is still made.

These days, in a small wave of enthusiasm, the making of natural indigo from plants has started again in various parts of India, with the best plants being grown in Bengal. It's an intelligent idea to grow indigo as it's a nitrogen fixer, is good for the soil, is non-polluting and eco-friendly. I visited Kolkata every year on my gastronomic tours, and a wonderful shop to buy hand-woven natural indigo clothes is Maku Textiles. There, Santanu Das makes ethical, sustainable textiles and believes in protecting the heritage of indigo growing and dying and hand-crafted textiles. Lots of ethical up-market Indian designers like Aneeth Arora of Péro and Chinar Farooqui of Injiri, both based in Rajasthan, also use natural indigo hand-woven cotton for their designs.

When I was in Ahmedabad in Gujarat (the textile centre of India), I sashayed up to the swish new Arvind-Indigo Museum. This space was created by textile company Arvind Ltd not only to chronicle the history of this magical colour but also to reinvent and extend it, giving it new relevance. It was their foray into denim (therefore indigo) that saved the company's fortunes. Here you see that indigo can be used not just for fabric like cotton but also brocade, pashmina, silk and canvas. In addition, they make indigo paintings, art objects and installations, using dyed concrete, wood, aluminium and enamel. Yes, unbelievably, aluminium can be infused with indigo dye, a process that has been patented.

As a fascinating aside, indigo is why our mothers used Bluo bags in the washing machine to keep white fabric bright. When you put a blue rinse with white, it becomes more white by disguising the yellowing, and it's also anti-fungal. It has blue iron salt (ferric ferrocyanide) in it and for centuries has been added to rinse water to make us think the white garment is whiter than it really is: it's the old UV light trick I mentioned earlier — blue light tricks the eye into thinking it's seeing white instead of dingy yellow. Unsurprisingly, white is fantastic paired with indigo.

Green

Any preponderance in green of yellow or blue introduces a corresponding activity and changes the inner appeal. The green keeps its characteristic equanimity and restfulness, the former increasing with the inclination to lightness, the latter with the inclination to depth. In music, the absolute green is represented by the placid, middle notes of a violin.
—Wassily Kandinsky

In the 2007 movie *Atonement*, Keira Knightley's character wore an electrifying emerald-green silk gown designed by Jacqueline Durran. It was inspired by 1930s' fashion, an era when clothing was freeing up, but went a lot further. It was clinging, wispy straps barely holding it on, and had a vertiginously plunging backline and a seemingly liquid train in the skirt. The whole thing was pulled together with a drape tied low around the hips. This is the sort of dress you could only look good in if you're 20. It was, though, unforgettable, with copies selling for over US$30,000.

In an interview with *Entertainment Weekly*, Durran explained that the director had asked for green but hadn't specified what type: 'We went to all of the fabric shops in London and gathered up all of the green fabrics we could find, in different qualities — some were more transparent; some were darker; some were lighter; some were heavier — and we played around on this table with all of these green samples. In the end, we put together three fabrics — one on top of the other — which were two transparents and one solid.' Having selected this intense composite colour created by this layering, they gave it to a dyer to reproduce with the silk they had decided on, which was originally white. Asked why that was the right colour, Durran replied: 'I don't know', and yet there's no denying it's the one piece of clothing most people will recall years after having watched the movie. The takeaway lesson from this is that it's all about finding the right green for you.

The trick with wearing green is you have to choose a hue that suits your skin and hair colour — emerald, forest, chartreuse, lime, grey-green, apple, pistachio. Rosemary green is of course ravishing with its purple/blue flowers. Redheads can get away with any hue of green, but blondes might choose rice-paddy green — bright and clear with hints of gold in the palette. Chartreuse (between green and yellow) is most often used in interior decorating but looks stunning on pale women with brown or dark red hair, if you have the guts to wear it; it could be paired with mustard, burgundy and ox-blood red.

Green calms and is restful and undemanding on the eye because it is reminiscent of the sea, sweet peas, fields and nature — it is a harmonious colour for the bedroom and a very flattering colour to wear, especially for redheads. In spite of it not being used very much in fashion, it is sophisticated and elegant. Even if you live in a city apartment, if you can see green trees you are happier and more balanced. On the down side, green represents jealousy and betrayal. This comes from the ancient Greeks, who believed jealousy was the result of too much bile, thus making the skin green. This colour's strange history continued on into the Middle Ages, where it was against the law to make the colour, not only because of its association with poison, badness and general volatility, but also because it was illegal to mix two colours to make a new colour: to make green you had to mix woad (blue) with weld (yellow). *And* it was hard to make a good green (sometimes the arsenic in the dyes poisoned people). During the times of the Crusades in the twelfth century, green was ostracised by Christians because it was a colour associated with Muslims. Even in Shakespearean times, it was thought to be bad luck to wear green on stage. It has always been regarded as tricky to paint with because of its lack of clarity and fickleness, unlike red, which had more staying power.

Some greens have wonderful stories attached to them; for instance, the iconic absinthe, which was not only a colour but a drink. This pear-coloured drink made from herbs, aniseed and alcohol was featured — and consumed — by many a famous painter in the nineteenth century, hence the expression 'l'heure verte' or 'the green hour', which saw half of Paris reeking of 'herbs' at the precise hour of 6pm. It's a gorgeous colour, though. Celadon is another unusual green — luminous pale blue/grey/jade — and very flattering to wear. It was discovered in China in the tenth century and used as a pottery glaze. Some people find celadon rather boring and dull, but if you look on a good colour chart you will see how beautiful it is.

Because my mother was Irish, I threw myself into Irish dancing as a child, a realm overdosed with emerald green. It didn't necessarily suit my pale, freckled skin and black hair, but what did I care? I was a

kid who just loved any ostentatious outfit, no matter what it was, and getting all dressed up in green, white and orange with green bows in my hair was the least I could do to contribute to the cultural life of 1950s New Zealand.

I wore a glorious green dress on the cover of the *Australian Women's Weekly* in 2018. It is similar to the ones I used to make for the Cook Street market but better quality and I paid a lot of money for it. It is green hand-screened silk covered in birds and flowers and made in Jaipur by Banjanan. It is as light as a cloud and you could wear it to the beach with jandals or wear it to the opera in gold high-heel sandals. When this cover came out I could have sold a hundred copies of that long, flowing green dress.

Diggi, my manager in India, had told me about Banjanan, which is based in Jaipur and New York. They had a workshop in Jaipur where, if you had an introduction, you could buy clothes. So, when I was in the city, I jumped in a tuk-tuk and spent such a long time trying to find this place that I was in a very bad mood when I finally got there, and almost bit the lovely designer's head off. Caroline Weller graciously let me try on absolutely everything in her beautiful studio in a beautiful house on an up-market street. The brand is sustainable, all hand printed and embroidered in Jaipur and bursting with colour.

I have worn a number of greens over the years. For instance, the Marni green-striped silk dress that I found in the Scotties designer recycle department I wore in 2013 to Government House in Auckland when I received the New Zealand Order of Merit from the then-Governor General Jerry Mateparae. I wore it with chandelier earrings from Kolkata. Then there's the Lisa Ho sundress I bought while in Sydney filming *Taste Takes Off*. I paired it with a very old Scotties cardie and Marni necklace for the Sydney shoot on Persian pashmak (candy floss) and luxury food products like pistachios, nougat, saffron and rose petals. We did a lot of research on this company Pariya, set the interview up imagining rose petals falling from the heavens on us, flew to Sydney, I got all dressed up in my new green sundress and three kilos of make-up, the day came and . . . the interviewee suddenly cancelled the shoot. I've forgotten why exactly but I seem to remember something about our high-rating show not being glamorous or up-market enough for them. We were used to people begging to be filmed because it always had dramatic results in terms of sales, and no one had ever turned us down before, let alone said yes, wasted our money, then let us down. So every time I wear this dress, I think of candy floss and rose petals and feel the overwhelming need for Iranian-style rice. They steam the rice in layers interspersed with rose petals, nuts, currants and herbs. The base of the pot has butter and oil in it, so the bottom later of the rice is always crunchy and golden. The Iranians are the best rice cooks in the world, but these people, while classy, were ill-bred. Still, I got a green Lisa Ho dress out of it.

There is, you'll be pleased to know, a green wave currently taking over the trending colours: very pale mint suits (lots of this colour is being used), lime-green leisure pants and tops (if you don't mind looking like a tennis ball), neon-green stilettos, emerald green paired with purple, emerald green paired with turquoise (surprisingly cool), green coats covered in large brown flowers.

Black

I don't need to discuss black at length because it is not a colour. It is the absence of light. The only women who ever looked good in black were Coco Chanel and Audrey Hepburn in her iconic black Hubert de Givenchy dress in *Breakfast at Tiffany's*. To me black symbolises fear, bleakness and self-loathing. By now you'll have gathered why: the dictates of my school-teacher nuns to wear all black except underneath was ground into me, and for all I know into a few rugby players as well. I reacted, though, by not letting the habitual habit become my habit. But for good aesthetic reasons as well: most people don't suit black. I'm sorry, but it is a dark, cold and harsh non-colour, so if you are mellow, fair or pale in your colouring, don't go near it — it is ageing and, like a vampire, drains the colour out of your skin. Don't believe me? Look at your average vampire.

The reason it is unflattering and ageing is because it widens and drops the jawline, making you look like you have a double chin, and causes shadows to pool in lines and crevices under your eyes when worn too close to your face — think of all those Greek widows. In black, the older you are, the paler you'll appear; the lighter your hair, the worse you'll look. One of the many fashion lies we're force-fed is: 'black makes you look thinner'. I've said it before and I'll say it again fat people in black just look dowdy and, surprise, surprise, like fat people in black. Fat people in colour look happy and fabulous. Why do women wear it? — there really are not that many funerals in the world. To use my food meets fashion philosophy, there's a good reason very few edible items are black: because they just don't attract the eye.

Okay, just as liquorice, squid-ink pasta and black beans are worth paying attention to, some people can wear black, but just make sure it isn't wearing you. At least wear red lipstick or bright earrings or a bright scarf to mitigate the damage.

The true secret to any possibility of looking good in black is the fabric — all fabrics that reflect light, like alpaca or mohair, are attractive. However, some fabrics hold black dye better than others and some textures look better together than others. Wool is always a deeper black than cotton and holds dye for longer, thus fading more slowly. Black cotton fades, so it's best to wear a cotton/Lycra mix because the Lycra holds the dye. To retain the dye, black garments must be washed separately on the cold setting with a cup of vinegar thrown in.

Oddly enough, men look good in black, especially all black — shirt, jacket, trousers, socks, shoes — this is inexplicably exciting, dangerous (Johnny Cash), sexy and masculine. Men wearing all black is so hot it's unbearable, and yet not that many do, excepting Italian gangsters and gypsy jazz musicians. Men can get it wrong, though, and end up looking like the parish priest where everything's a bit faded and tawdry.

Having been so damning about wearing black, the second-ever Marni piece I bought was a black tunic (who says I have to follow my rules, though maybe the red on it is what really attracted me?). I found it on sale in Harrods in London in 2005. I have worn it endlessly, and it still looks exactly the same as the day I bought it. It is perfect with black tights, leggings or Capri pants and high heels. When I purchased it, I was in the capital with Television New Zealand shooting a story on tea — Twinings tea to be precise. We had a fantastic time endlessly tasting exotic teas, like dragon whale, oolong, brick and jasmine. I invented three really weird recipes for this show: green-tea soup with rice and tuna; pear salad with green-tea and mustard dressing; and berry and green-tea granita.

On the other hand, using tea to smoke meat, fish or vegetables instead of wood or coal is really delicious. And here black can hold its own, as black tea generally has the strongest flavour, and teas like smoky Lapsang souchong or the bergamot-oil infused Earl Grey will provide extra notes. You put the tea in the bottom of a wok, place on top a cake rack and lay the food on it. Cover with tin foil and smoke for 20 minutes, while you don a little black number. And as we're on tea . . . everyone knows that women are like teabags — it's only when you put them in hot water that you realise how strong they are.

Black Marni, though you could also call it red and white: not many people can get away with just black; even a blackbird has a yellow beak and eye-ring.

White Indian dress from Péro and a white and blue blouse from Brigitte Singh.

White

And all woke earlier for the unaccustomed brightness. Of the winter dawning, the strange unheavenly glare: The eye marvelled — marvelled at the dazzling whiteness . . .
—Robert Bridges

White instils reverence in the human heart. White doves symbolise peace and love, though do be aware that in some Asian countries white is the colour of mourning clothes. Like blue, it is another of the very old colours. A symbol of purity, chastity and serenity, white is cool, calm, sophisticated and exceptionally flattering, as it reflects light onto the face — think of the translucence of crisp cotton, the delicate beauty of lace, the holiday feel of crushed white linen. Like black, white is not really a colour, but it suits absolutely everyone and goes with absolutely any other colour — especially black. If pure white turns you into a ghost, try cream or off-white; there will be a shade that works with your skin tone.

It's not a terribly easy dye to make — you have to start with a white pigment (like lead white) or titanium white (which is what makes toothpaste, pills and tennis court white), otherwise you are just going towards black. It's hard to mix because of the way our brains activate light. Lead white is a thick lead carbonate with a crystalline molecular composition, first discovered in Anatolia in 2300BC, and it is still being made to this day. It was dangerous, though, and could result in lead poisoning — inspired by Elizabeth I, women indeed suffered for the sake of beauty, using a white lead-based foundation that slowly ate into their skin and eventually killed them. Even by the eighteenth century, they were still using it to look fashionably pale and to cover smallpox scars.

With the discovery of silver in Bolivia in 1545, the colour, a relative of white, became fashionable for clothing. It is associated with the moon and is often used in evening wear, signalling elegance, grace and silkiness. I met the first man I almost married while wearing a silver satin dress cut on the cross. He still talks about that dress. And while on the subject of (almost) marriage, the obvious white clothing that comes to most minds is the wedding dress. However, brides mostly chose coloured fabrics, often woven or embroidered with flowers, until Queen Victoria wore an ivory satin wedding dress in 1840. Ever since, Western brides have worn white.

Having worn white for so many years, as a baby, first communion supplicant, debutante, nurse and chef, I have trouble wearing it now due to post-traumatic white stress disorder. Incidentally, the reason why these uniforms are white is so they show the dirt. The message is: don't even think about being a slob, because cleanliness is next to godliness. Possibly doctors first wore white coats in the nineteenth century because at that time white wash, made from a simple mixture of lime and salt, was used to paint buildings and streets to disinfect them from the plague. The expression 'to whitewash', to conceal disagreeable facts, comes from this period.

Despite the white stress disorder, I like the 'dirty' white of old fabric, lime wash and silver. Shades of white can be mixed, but they have to be in the same family of shades otherwise one of the whites can look discoloured. It was always assumed that if you wore white clothes or had white furnishings, you were so rich that you could afford cleaners, but these days people don't care if white is dirty — it seems to denote you are bohemian and beyond such bourgeois concerns as appearance. Personally, I think a crumpled white linen suit on a man is rather sexy and *je ne sais quoi*. I mean, you wouldn't say no to a man in crumpled white linen, would you? Conversely, I'm suspicious of a man in a perfectly pressed suit — seems to have a whiff of sanctimoniousness about it.

I also like the bone white of ancient Greek and Roman ruins, but guess what? All these buildings and statues were originally painted in bright colours that have since faded or washed away. The sculptor Rodin said he didn't believe it and deep down inside he just knew they had always been white. He was wrong. The lesson from this is: don't cling to something because you think it's always been that way and is the only

My white Pleats Please blouse, with dress knotted for storage, earrings and subtle little shoes.

way to go. A bone-white statue is beautiful, but so is a multi-coloured one. A white wedding dress is beautiful, but so might be a harlequin one.

In the South of France where I live for half the year, the women have a pronounced white, dove, clotted-cream tendency. In their eyes, if you're older, you have to keep it simple and stylish — no flounces, frills or dressing like a baby. If you're younger, you can do what you like. I guess that makes me young.

I still wear a pretty white Pleats Please blouse that I bought in 1999. It is delicate and fine and pleated across rather than up and down, which gives it a floaty, elegant feel and appearance when worn. I matched it with a long straight yellow skirt from Scotties and a black corset for a *Taste New Zealand* photoshoot. I was eating grapes. I don't know how I got away with it.

The most famous ivory dress ever worn was on Marilyn Monroe in the 1955 film *The Seven Year Itch*, and was designed by William Travilla. The photo of her standing over a subway grating with the wind blowing up her cocktail dress is one of the most archetypal images of the twentieth century. This dress just screams sexuality, abandon and thrilling disruption. It is made of pleated fabric with a halter top, and has a demure little tie around the waist and a full skirt — arms, neck and chest, shoulders and back exposed. In a gossipy aside, it was rumoured that Travilla didn't design it at all but bought it off the rack (which he denied). Debbie Reynolds acquired it for her collection, and subsequently on 18 June 2011 it was put up for auction — it went for US$4.6 million, plus 1 million commission. The most famous white dress ever worn by me is my debutante dress in 1966. I did not look like Marilyn, and it is not for sale.

Orange

Orange used to be called *ġeolurēad* in Old English, or yellow-red; apparently the first time anyone heard the word orange as a colour name was in the sixteenth century when Elizabeth of York described some fabric she had bought as 'orenge'. Kandinsky said that orange was 'red brought nearer to humanity by yellow'. Orange is a determined, vibrant colour, and you have to be confident to wear it — it is on the opposite side of the colour wheel to blue, which is shy. Some prison uniforms are orange, traffic lights are orange, terror alerts are orange, and did you know airplane black boxes are in fact orange?

One of my evening dresses is orange — it means watch out, close the front door, tornado coming! Near where I live in the South of France is the city of Orange, not so named for the colour or the fruit, as I thought, but after Arausio, a Gaulish god. Orange — the colour, that is — has had its moments in art and fashion, but has never been big permanently, like its glamorous sister red, because it's difficult, doesn't suit everyone and is considered to be too brilliant to be elegant. It's what you call a slow grower, but it is used much more than it used to be. It appeared a lot in Art Deco fashion illustrations in the 1920s and was madly popular in the 60s and 70s (I know because I was there and there was a lot of criminality involving brown and orange in kitchens). It gained currency at the time as it was considered bohemian and counter-culture; also it was (and still is) the robe colour of both Buddhist monks and Hare Krishnas.

This sequinned tangerine-coloured dress by Trelise Cooper gives me a new appreciation for wearing chainmail (photo © Lawrence Smith).

Orange is the colour of the fabulous Hermès packaging, though only by chance. After the war, due to shortages, they had to take the only colour available, which was orange, and they turned it into the most iconic colour in the luxury fashion business, signifying sophistication and class. There is a similar shade called Dutch orange, which is a sunny mandarin, associated with the Dutch House of Orange, of which today's royal family are descendants. The Dutch were rather obsessed by the colour and even made carrots orange — originally they were purple or yellow, and in the late sixteenth century they selectively bred the other colours out of them. In the seventeenth century, the Dutch even tried to call New York 'New Orange', until they lost the city to the English. The Dutch still think they are orange and their sporting events look like a sea of Jaffas.

A very beautiful sister of orange is the colour of the ancient saffron spice, long used to flavour food and dye fabric (another food and fashion connection). It is the most expensive spice in the world — 5 grams costs NZ$43. Saffron is a heavenly, intense, lasciviously crimson-gold colour and comes from the *Crocus sativus* flower. The best is grown in Iran, though we grow a bit in New Zealand. It is very easy to fake saffron with turmeric, dyed corn silk, dyed coconut threads and safflower, but there's a way to be sure you're buying the real thing: smell it. It should smell musky and sweet. Another way is to soak it in water; initially the water will turn yellow, then gradually dark orange. The most reliable way is to be able to recognise what a saffron stigma looks like — it is thin and trumpet shaped. Never buy saffron if it is cheap and never buy it in powder form, which is too easy to adulterate.

Eye-popping saffron-coloured clothes are very flattering on dark skin tones and go brilliantly contrasted with true blue or chive green. It is considered to be a classic spring colour in fashion, but is also delicious in winter. My mother made me a burnt-orange duffle coat when I was a teenager. I think I was the only girl in 1966 New Zealand with an orange coat — she was so ahead of her time. She herself had a burnt-orange suit my father brought back from America. She wore it with

Pleats Please strikes again, but is it a pink dress with orange and blue stripes, or orange with pink and blue or blue with . . .?

brown patent-leather high heels and orange gloves and called the colour 'hot orange peel'. In 2019 Kendall Jenner wore a giant orange feathered gown to the Met Gala — designed by Versace. She looked like a tall Guianan cock that had just flown in from Brazil.

When you need a bit of va-va-voom, Trelise Cooper is definitely the way to go. I bought one of her orange sequin dresses for my one-woman shows, which toured New Zealand. The orange silk is embedded with millions of sequins, and to wear it was a labour of love, as it weighs 2.5 kilos. I teamed it with chandelier earrings and silver studded shoes. When I was awarded the New Zealand Order of Merit in 2012, I wore it to the after party. My sister Keriann had come up from Hawke's Bay with a huge cake she had made. We hid it under the bed in the office and then forgot about it. This is what happens when you wear 2.5 kilos of sequins.

Pink

The day French men started wearing elegant pink shirts was the day I started loving pink. I have to say I'm not wild about pink in fashion unless it is a strong hot pink (which is very Indian — I called my book on Indian food *Hot Pink Spice Saga*). After all, pink is just faded red, and the word pink didn't even exist till the late seventeenth century. I do though love it in interior décor — I have pink couches in both New Zealand and France and pink flowery cushions, and in the past I have had pink walls. Whole restaurants in India are painted in this colour; in fact, the whole city of Jaipur is pink. Diana Vreeland called it the 'the navy-blue of India'. I wanted to paint the exterior of my house in Uzès dark pink, but the heritage association wouldn't let me. Maybe they were right.

Pink is flattering, warm, pretty, restful and particularly good on olive skins. Pale-skinned women like me don't look desperately fabulous in this colour because it's just too similar to the pink tones in our skin, but then again it really depends on the shade. Baby pink is for babies, okay — or is that me having a feminist backlash moment against dated sexism? The vivacious and cultured Madame de Pompadour, mentioned above, *loved* pink to the point where she had Sèvres porcelain coloured bright pink. Marilyn Monroe wore an explosive shocking pink dress designed by William Travilla in the 1953 film *Gentlemen Prefer Blondes*. It was satin, floor length, strapless and had a huge bow stuffed with ostrich feathers at the waist. Barbie doll had a version made for her, and Madonna wore a copy for her 'Material Girl' video.

Pink has many shades — fuchsia, fluorescent (highlighter pen), puce (actually a dark muddy pink), amaranth, pastel, magenta, rose, to name a few — and many enchanting stories to go with them. Fuchsia is a bright, blue-based pink named after the flower by Mr Plumier in 1703. He called it fuchsia after his botanist hero, a Mr Fuchs. 'Shocking pink' hit the planet in the form of a large pink diamond. It was bought by twentieth-century bad girl Daisy Fellowes, who showed it to her friend the surrealist designer Elsa Schiaparelli. Schiaparelli described the colour as 'Bright, impossible, impudent, becoming, life-giving . . . a shocking colour, pure and undiluted'. She used it in her perfume label, in clothes and in interior décor. My favourite French designer, Christian Lacroix, uses it a lot with red — it is so good on olive-skinned Provençal women.

Christian Lacroix gave up his unprofitable fashion business for opera costume design, where he could indulge his love of excess and trademark unusual colour combinations, including red and pink.

Amaranth is a dusty grape colour, named after the flowers of the plant, which has edible leaves. Magenta is a pinkish purple that is exactly half way between red and blue. It's called the 'colour that doesn't exist'— it isn't on the light spectrum so doesn't have a wavelength for humans to see, but we can see it anyway. What happens is the brain makes magenta visible by filling in a space that works for the eye, which sort of averages other wavelengths out and fiddles around till it can see magenta.

In 2008, when in Paris, I couldn't resist a pink, black and white check cotton dress, and I think I wore it for a week straight, I was so enchanted with it. I still wear it from time to time. The great thing about it is that it is loose, so you can be fat or thin, still wear it and it still looks great. I went out to dinner with friends wearing it and made the mistake of complaining about the food when the waiter asked me what I thought of the meal. In response, he flew into an abusive rage and we had to pay quickly and leave. This is what happens when you wear a strong pink. It's a colour that aggravates you and makes you tell the truth, when it is clearly not in your interest. I have got into trouble for opening my mouth many times in my life, starting when I was five, and this is the only time it happened because I was wearing pink.

Mauve

The colour mauve is not desperately fashionable now, but there was a time when rich ladies were drenched in it — it was hard to find someone wearing another colour. Mauve is not purple, which I regard as the colour of clotted blood. Purple dye was originally created in 48BC and first worn by Cleopatra. It was made from the glandular secretion of two types of sea snails and called Tyrian purple. It was very expensive to produce (every snail had to be caught by hand) and tricky to get the dye to stay in the cloth. Caesar had a toga dyed in Tyrian purple and forbade anyone else from wearing the colour. So while purple was a natural dye, mauve, which came along centuries after, is chemical and man-made.

Young scientist William Perkin discovered mauve by accident in 1856 when he was trying to make quinine from coal tar in order to treat malaria. He never succeeded in extracting quinine, but he persisted with his mauve discovery and ended up making the expensive colour affordable. Perkin had concocted a dark oily sludge in his efforts to make quinine. Because he loved art and colour, he took a piece of silk cloth and dipped it in the muck and it turned the most exquisite shade of pale violet. Soon after, the wife of Napoleon III, Empress Eugénie, realised in a moment of profound narcissism that mauve was the exact colour of her eyes. Enough said. It was only months later that Queen Victoria attended her daughter's wedding resplendent in mauve and silk *moiré*. Immediately London was affected with 'mauve measles', but, like all infectious diseases, it eventually burnt out and subsequently got a bad reputation as a colour for menopausal ladies — *quelle horreur*.

Designer Neil Munro Roger, otherwise known as Bunny, co-invented the best pants in the world in the 1960s: Capri pants (cropped and slim-fitting — just the sight of them makes you want to have cocktails at sunset on the island of Capri). Audrey Hepburn and Brigitte Bardot loved them, but I digress . . . mauve. Bunny, an incorrigible dandy, dressed himself head to foot in mauve for his seventieth birthday, giving rise to a new moniker: the *eminence mauve*.

The secret to wearing mauve is the other colours you put with it. If you wear all mauve you look jaundiced, but if you put yellow or orange with it, the colour comes back into your face. I have some Designer's Guild mauve silk furnishing fabric embroidered with gold, turquoise, purple and green flowers, so I see no reason why those colours wouldn't go together in an outfit. I'm thinking mauve Capri pants with a turquoise sweater and turquoise pointy kitten-heel shoes or a chocolate-brown dress with a lightly quilted mauve summer coat.

Bring back mauve, I say.

Marni's refined but dramatic mix of patterns and colours are my kind of style.

CHAPTER 6

Mixing and Matching

The monochromatic look

With colour still in mind, let's move on to combining elements. Perhaps the most obvious is the monochromatic look: black and white are good together and they both have many variations in tone. Black can embrace deep midnight blue-black, burnt brown-black and charcoal. Van Gogh proclaimed that he could identify 27 different shades of black. Meanwhile, there are dozens of shades of white, such as cream, eggshell, ivory, latte (otherwise known as beige), bone, grey, alabaster and vanilla. In the paintings of European masters, there are often luminescent white lace collars on black velvet gowns and glimpses of black-and-white checkerboard pattern floors. I love black-and-white check floors — they are classic and go with every

Over 20 years old, this black-and-white Marni dress is as striking as ever.

décor. Just like a black-and-white outfit — you can put any jewellery or shoes with it.

The black-and-white fashion photos taken by Willy Maywald in the 50s and 60s are so much more moving and nuanced than colour photos. A woman in a black trouser suit with a cool white shirt is simple but show-stopping — kind of sucks anything boring right out of the room. Black-and-white stripes, checks and dots are wonderful and don't have to be strict black and white, which makes it so much more interesting for dressing. My favourite black-and-white dress is the eighteenth-century Gothic striped gown worn by Christina Ricci in the film *Sleepy Hollow*. It just has everything to keep you wide awake for the next three weeks. It has a bustle, it's voluminous and has vertical, horizontal, diagonal, flouncy and frilly stripes — have a look online. It was designed by Colleen Atwood and copied many times.

My Marni black-and-white dress is the most enthralling dress because it is voluminous, complex and entirely held together with one tie on the side. It has no zip and no buttons. If you pull the tie the dress falls open. I take particular pleasure in telling men that at parties. I think I bought it in the recycle department at Scotties in Auckland in 1997.

On one occasion I wore it on the cover of *Metro* magazine for a story on the top cooks in New Zealand. The cover photo featured us all — we were told to wear black or white and not to smile — that was the first cue I missed. The second cue I missed was that the journalist insisted on interviewing us separately even though we all knew each other. Ray McVinnie advised me not to do it as the writer would have his own agenda. 'He's just a journalist, Ray,' I said, 'it's just a story about good cooking, what could he possibly do?' When the story came out we were all horrified as we felt he had put us down and pitted us unknowingly against each other. I was in Morocco when the story came out and got tearful phone calls from various other participants, apologising for the way he had misconstrued our comments about each other. Every time I wear this dramatic dress, I remember that this journalist almost ruined my friendship with one of the cooks, and I think of that photo of us all in stunning black-and-white clothes looking like a bunch of hard-nosed bitches. Nevertheless, this dress has also given me a lot of pleasure and I love the big, wide neckline and secret folds everywhere.

An alternative way to mix colour is what the French call 'camaïeu'. It is the monochrome technique of wearing different shades of the *same* colour. Not all exactly the same colour, though, because that just means either you think you are a bridesmaid or you think you're the Queen. Most people would do this in black tones or a mix of whites and creams, but I think it's very wonderful in blue or red: such as maroon cigarette pants, madder red shirt, fuchsia pink coat, gold high heels and big gold necklace; or a long, gathered cornflower blue skirt, tight indigo blue sweater, vintage powder-blue Doc Martens and Indian earrings.

Pattern

I am a big fan of mixing patterns that strictly speaking shouldn't go together. Some people can't even get their heads around one single pattern and, out of terminal terror of looking inappropriate, always wear block colours. The desire not to look like your grandmother's dining-room chairs overrides all experimentation. Or, as my publisher

When it comes to combining patterns, take inspiration from the varied decoration of Moroccan palaces. Bottom left: a Marni dress with Andraab Indian hand-embroidered scarf; bottom right: pants by Chloé Stora (a French label) and blouse by Attic and Barn (from Italy).

puts it, the fear of looking like the carpet under the Christmas tree. So why would you do it? Because it's the closest you'll get to art if you're not an artist. In Moroccan palaces, the floors, walls and ceilings are often decorated in an exuberant mix of patterns and textures (tiles, carved stone, painted wood, etc.) that we're simply not used to back home, but a slight adjustment to your outlook and you'll realise multiple patterns can party together with style.

Making the most of patterns in clothing has a long history, but for a not-so-distant example let's go to the Great Depression in America. In those days flour was always sold in cotton sacks. Cotton was hard to come by, and no one had any money anyway, so the women washed out the labels on the flour sacks and made clothes out of them. The manufacturers saw what these women were doing and began printing the flour sacks with gorgeous designs — really wonderful prints of flowers, dancing girls, little animals and birds, teapots, geometric designs. Sometimes the sacks even had instructions and patterns printed on them for making toys, diapers, dishcloths, tea cosies and tea towels. You just washed the instructions out and got sewing, combining whatever prints you had. Which brings me to quilting.

When the sacks and clothes finally wore out, they were turned into patchwork quilts. Patchwork quilts were originally a recycling and repurposing technique, but they are also works of art: the bits and pieces of varied patterns are thoughtfully placed so they speak to each other. Strikingly different patterns might be included, but often a common colour will make the combination work. Borders or cross-hatching echoing or contrasting with the patterned colours pull the designs together. You can use the same principles when dressing, remembering to mix with style and compatibility.

To get an eye for what can work, it can help to look at garments that combine different elements successfully. One day in 2016, I was visiting a New Zealand friend in Chennai, the capital of Tamil Nadu, India. It was very hot and I was on my way to host a culinary tour in South India. Chennai is not exactly the most gorgeous city in the world, but it has of late become rather cosmopolitan, glamorous and luxurious. It is a big, noisy city — a whopping 426 square kilometres — on the spectacular Coromandel Coast on the Bay of Bengal and is the keeper of South Indian artistic and religious traditions. The Tamil film — and film-star — business is reaching fever pitch, and film stars often become politicians, which is why they are so silver-tongued and well dressed. It has areas known as Ville Noir (Tamil) and Ville Blanche (French), and is famous for its 'messes' or cheap, delicious food houses.

So my friend and I were doing a tour of the best clothing shops, like Good Earth, Nicobar and Amethyst, and ended up at the best of all, Collage. Collage stocks a range of luxury brands like Rasa, Eka and my favourite, Péro, which they have rows and rows of. Indians love wearing completely over-the-top wedding outfits covered in gold and jewels and heavy embroidery, and this shop is full of them. I can't stand them and always go straight to the natural, indigenous, cotton end of the shop.

My friend looks at all the beautiful Péro clothes, saris and shawls and says, 'No, you are not paying this amount of money for a dress, Peta, I forbid you.' I either touched or tried on everything, then chose a flowery reversible sari blouse and a yellow and navy-blue check dress. When I swished around the shop in this dress, stepping over the bodies of the sales assistants who had swooned with the fabulousness of me and the dress as one entity, my friend said, 'You are NOT buying that dress. It doesn't suit you or your body at all and just looks ridiculous.' So, against my better judgement, I passed on the dress and bought the blouse. Gentle reader, I could *not* stop thinking about that dress. I went back to my hotel on the beach, welcomed my gastronomads and next day left for the first stop of the tour: Pondicherry. I told my Indian manager Diggi the story of the yellow Péro dress. He has known me for years and shamelessly indulges my passion for clothes. There are many heroic instances where Diggi has gone beyond the call of duty for textiles, scarves, silks and saris for me, but he was to outdo himself in the case of the yellow dress.

After a few days in Pondicherry, the tour moved on to Madurai. I'm

The Péro dress that nearly escaped my clutches is a riot of elements and yet everything works together in restrained harmony.

still banging on about the dress and how much I regret not buying it. I'm sitting on the bed in my lovely room with the private swimming pool when there's a knock at the door. It's pouring with rain. I open the door and Diggi is standing there holding up the dress on a hanger. I scream and scream. This is why I love Diggi. He had bought the dress, then arranged for one of his drivers to pick it up and drive over seven hours from Chennai to Madurai to deliver it to my door.

The dress has small navy-blue checks on the bodice, with an egg-yolk yellow, very fine cotton lawn gathered skirt. The fabric is block-printed by hand with tiny flowers. All along the seam between the bodice and the rest of the dress are handmade colourful cotton tassels, and the hem of the dress is scalloped in red thread. The tiny brass buttons are handmade, and the neck is embroidered with red French knots. The whole thing is lined with a very fine white muslin slip terminating in a large navy-blue check border. I have featured this dress in magazine photoshoots, worn it endlessly for years, from the jungles of Goa to the bright lights of the Champs-Élysées, and had countless people complimenting it and asking me who made it.

Stripes and spots offer a less challenging way of wearing pattern,

I went for stripes and spots for the cover of my memoir *Never Put All Your Eggs in One Bastard.*

though they may be worn on the same day. I love stripes mixed with prints, for example, a Jean Paul Gaultier sailor T-shirt with a short, large-flower print skirt, tassel earrings and espadrilles. Stripes and flowers are made for each other. In Paris it is not uncommon for half your friends to turn up at the bar in stripes, so you'd be forgiven for thinking you were at a Maritime Union meeting. A stripe and a spot just make everyone look adorable and fresh — look at Minnie Mouse. Navy-blue and white dotty shorts are good with a navy-blue and white striped shirt, a bomber jacket and blood red lipstick. I have a cotton sari block-printed by hand with huge dark pink spots that I wear with a tiny-flowered blouse, and they are very beautiful together.

The sari

Talking of saris, when it comes to successful mixing and matching, you can't go past India. Not only do you see the most stunning colour combinations but also a glorious abundance of patterns. Saris can be plain, but they also come in intricate patterns, edged by borders in different designs. Rajasthanis in particular wear startlingly riotous and colourful clothes — you can see a Rajasthani woman for miles because of the electric colours of her outfit, and a man by his huge colourful turban. Their clothing design is not there just for beauty. Traditionally they have a social message woven into them, indicating the village and sub-group you belong to, your marital status, societal position, status, occupation, religion and gender. For example, you can tell by a man's turban or *pagdi* if he is single, widowed, rich, poor. This is achieved by a system of knotting, shape, material, colour and pattern. In some places, such as Balotra in the Barmer district, south-west of Jodhpur, if you are married your skirt or *ghagra* will have a red border with yellow piping stitched to the bottom of it. If your husband dies the red trim goes. They have mixing and matching down to a fine art of communication.

At one point Trelise Cooper was buying a lot of Indian fabrics. I acquired one of her skirts that was fairly outrageous (but not by Indian standards), a cross between a flamenco dress and a Rajasthani *ghagra* skirt. The outside is machine-embroidered lime (almost yellow) silk, while the lining is fuchsia silk with five layers of frills. Originally a corner of the frilly lining was pinned up with a brooch onto the skirt to expose the frills. I wore it with a white top, big earrings and high heels to do an interview on the beach in Nelson. There was absolutely no reason to dress like an Indian Christmas tree but I just felt like it.

I have been visiting India since 2005, and although I admired the beauty and gracefulness of the sari on Indian women, initially it never entered my head to wrap myself in one. I just thought it was too outlandish on a white woman with freckles and red hair, and also too hot and cumbersome in the heat. How wrong I was, but more of that later. I had first been tempted to go the whole hog and wear a sari (but didn't quite make it) when I was introduced to Shamlu Dudeja, who specialises in handmade *kantha* work (which I write about later). In her very elegant home in Kolkata, she has a shop full of fabulous shawls, clothes and saris, all done in a tiny running-stitch embroidery called *nakshi kantha*.

Then I was introduced to Parama Ghosh, an indie designer in Kolkata. She has a law degree but decided she wanted just a bit more

This silk skirt couldn't be by anyone but Trelise Cooper.

fun and creativity in her life so now she makes saris, blouses, *dupattas* (shawls) and all sorts of things. Her designs are completely Kolkata and unafraid of bold designs and mixing and matching, with whimsical appliqué, embroidered flowers, Kolkata yellow cabs, rickshaws, motifs from film, art (Frida Kahlo), songs and literature all featuring. Parama is inspired by anything that is going on around her — weather, bridges, temples. She uses different textiles but mostly cotton, specifically Dhaniakhali cotton, made by handloom in a district of West Bengal and famous for its softness because of its 100 x 100 thread-count. It was in the presence of Parama's unusual saris when I visited her workshop that I broke down and bought a *choli* (blouse) with a taxi on it and a fine grey sari with bright red hibiscus flowers embroidered all over it. I thought I'd wear it at least once for fun. I first paired it with one of my embroidered Péro blouses, and now I use every opportunity I can to wear it — I had no idea how light and comfortable a sari is to wear and how elegant and cool you feel walking in it. One thing you have to remember once you have learned how to wrap at least 6 metres of fabric around yourself, is to wear a long petticoat under it, as they are often transparent and although there are many folds at the front there is only one thin layer at the back.

The sari originated in the Indus Valley Civilisation around 2800BC. Originally worn without a blouse or what they called a chest band, gradually a blouse was added for modesty. When folding it into place, you can reveal more or less flesh, according to your desires. It is possible to be very revealing of the midriff and belly button, complemented by a very tight *choli* (top), just as it is possible to show nothing and even cover your head and face with it, wearing a *choli* to the waist or, as in Rajasthan, a double *choli* split at the waist, to the hips. The sari can be wrapped in many ways, but usually you leave a long piece at the end to cover the chest and fall over the shoulder (this bit is called the *pallu*). Brides traditionally wear red saris and enough jewellery to sink a ship. My prediction is that if all fashion and clothing disappeared in the future and only one thing survived, it would be the sari.

How could I stop at buying just the one sari when I realised how wonderful they are to wear? This dotty beauty is teamed with a Péro blouse, the dot colour being echoed in the Péro roses, and the earrings and bangles echoing the abundance of the flower print.

Walking Art

If you really want to look like walking art, though, think about Yves Saint Laurent's famous art dresses inspired by the abstract geometric painter Mondrian — they came out in 1965 and he sold millions of them. The dresses were straight up and down so as not to interfere with the design. He was also inspired by Picasso's colourful abstracts, and in 1966 he interpreted Andy Warhol in pop art dresses. In fact, Yves Saint Laurent created a lot of the styles we now take for granted: tuxedos for women, tights, the pantsuit, trench coats. In the summer 2020 collection, Marni did gorgeous Matisse-inspired shirts and tops.

One of my prettiest Pleats Please dresses is a straight up and down, knee-length dress, very cleverly cut so that, once you put it on, the top part folds back into a collar that falls around the neck and over the shoulders. It is multi-coloured, with a predominance of blues and yellows and is printed with Japanese scenes, like a bus full of children, pagodas, flowers, trees, Buddhas, etc. When wearing it, you feel like a storybook and people can read you as you walk by. The fabulous thing about having lots of Pleats Please pieces, apart from the fact that they last forever, is that all the tops, jackets, pants, skirts and dresses can mix and match with each other.

Another of Issey Miyake's highly visual pieces is a 1996 orange dress and pants outfit, which I bought on sale while filming in Hong Kong. It's very loud and fits into the 'all I ask is a statement' category. If you wear both the dress and the pants together, you can easily be mistaken for a walking radioactive broadcasting station. On the front it has a picture of Lakshmi the Hindu goddess of wealth, fortune, luxury, beauty, fertility, power and auspiciousness. Almost every time I wear it, an Indian will come up to me and say, 'Do you know who she is?' And every time I reply, 'Yes, I do, she's my best friend and I am patiently waiting for her to come through for me.'

My Pleats Please
statement piece.

What can I say, a woman can never have too many scarves, but look at how the pink draws the wildly different patterns together.

Mix it up

The good news is that anyone of any age can wear patterns and it just takes a bit of *je ne sais quoi* to wear a mix — try it, you'll like it. A few rules are helpful, such as mix similar tones together like primary colours, or else all pastels, or all dark shades. Find a unifying colour in your mix, like red, black, pink, etc. You can really jazz up a plain outfit by wearing standout shoes: think gold sandals, animal-print stilettoes, platform sneakers, red starfish slip-ons, shocking-pink patent-leather pumps, powder-blue Doc Martens. You can also use scarves and big earrings to add interest to a look. You'd be surprised what goes together. But you won't know till you try it on; for instance, I have a long, flowery green dress that I wear with a pink Chinese print coat, and a long red-and-white polka-dot dress that I pair with a red Indian *kantha* scarf.

My New York fashion-buyer friend Greta is the mistress of the layered look. She'll pair a dotty burgundy dress with Gabriela Hearst lace-up short boots, light check dust-coat, dark glasses, with maybe a Gucci scarf tied on her handbag. Another time she wore Gabriela Hearst ankle boots, tights, a short frilly summer Isabel Marant dress, a fitting coat that did up, another loose coat over it, which she left open, and a large Denis Colomb cashmere scarf. Wearing layers of Pleats Please is always easy because it's so light and clingy — pants, tunic, top, scarf. I like Iris Apfel's eclectic over-the-top layering: oversize glasses, dozens of bangles, gigantic necklaces, multiple layers of colour and mixed patterns. Whoever said less is more wasn't thinking of Iris.

Here again your personality counts. The assumed wisdom is that small people shouldn't overcomplicate their look, nor wear big prints, but if you have a big personality, big features (e.g. a big nose like mine) or are a maverick, you can do what you like and get away with it. An example of personality-driven dressing is the Mexican artist Frida Kahlo. Her distinctive style and mixing of bold colours and patterns developed in part to mask her legs, damaged by the polio she suffered when she was a child and from a subsequent accident. You can do this too. If you don't like your legs, you can make pants and long dresses your signature style. Frida called it 'making myself up'. If you don't fancy super-feminine clothes, go with the androgynous look. If you have lost your hair, this is your opportunity to wear fabulous wigs and headscarves, or go bald with pride and chunky earrings.

Here are some examples of mixed pattern looks:

- A striped T-shirt with patterned skirt or shorts.

- A combination of pants, shirt and blazer that are all the same two colours (eg black and white), but all different patterns, e.g. stripes, dots, block colour, plaid.

- A leopard-skin print dress with a striped cardigan.

- Tight flowery jeans with a dotty Provençal-style shirt.

- A paisley print dress with a cashmere Indian shawl (the patterns can be completely different, just pull the look together by choosing colours with similar tones).

- Plain pants with a striped top and flowery jacket.

- Checks, tartans and stripes are cute together.

- Double florals, e.g. a top with tiny flowers matched with a skirt with big flowers.

- A patterned T-shirt with words or graphic print on it tucked into a dotty ruffle skirt.

- Double dots, e.g. a little-dot pale turquoise blouse tucked into a large-dot tan skirt.

My silk *ao dai* from Chula — how could I not buy it?

CHAPTER 7

Fabrics

To avoid being overwhelming or boring, it's a good idea to pair different fabrics and textures, for example, a thick jumper with smooth pants, a single-colour wool sweater with a patterned black skirt, a silk T-shirt with jeans. How do you do that with a dress? That's where jewellery comes in, or a belt, a handbag, interesting footwear, perhaps a contrasting shawl, cardigan or jacket, maybe a hat. Fabric is all about feel: how it feels on you and how it makes the viewer feel.

In recent years there has been a proliferation of new materials used for clothing fabric. Some are sourced from sustainable, fast-growing, natural products, such as the light but strong bamboo, the versatile stinging nettle (which reminds me of the Hans Christian Andersen tale of the princess who had to knit shirts for her brothers from this plant), and the antibacterial and durable hemp (which was woven into textiles in ancient times and is now making a comeback). Then there's the leather lookalikes made from pineapple leaves or alternatively grown from mycelium, bio-based nylon, not to mention fibres derived from coffee grounds, banana peel and lotus stems, among other unexpected plant products.

Some innovative recycled materials are made from such things as polyester from discarded plastic bottles, as well as reclaimed waste and industrial materials. And some new materials come from chemical wizardry and nanotechnology, including faux fur developed from polymers, materials that will regulate the wearer's temperature, new ways of waterproofing, and fibre coatings that can clean garments through exposure to light by the introduction of minute strands of silver.

Most designer garments, though, especially those that you can pick up second-hand, are made using traditional textiles, which have been tested over time, derive from natural sources, feel fabulous on the skin, and are durable and sumptuous. Indeed, while modern scientists are only now celebrating the benefits of silver, including it in clothing fabric has an ancient heritage.

Varak — silver and gold leaf

We humans have been spinning for possibly 20,000 years and weaving fabric for about 10,000 years, and the most luxurious, ostentatious thing we have figured out to show status is to spin gold threads. Of course the original golden thread appeared in Greek mythology in the guise of a golden ram. A king who was suffering bad crops in his land was talked into sacrificing his two children by a first marriage to the god Zeus by his dastardly current wife. She said that was sure to improve the crop. The children were saved at the last moment by a golden-fleeced ram who swooped down and told them to get on his back and hold on tight. They all rose into the heavens and escaped death. In gratitude the king later sacrificed the ram (which doesn't seem fair) and hung his golden fleece on a branch to dry.

I was in Kolkata in India with my friend Jane in 2014. I was hosting my culinary tour there and Jane was filming it to make a video for my website. Because she is very clever, Jane also produced my *Peta Live* one-woman stage show for many years. On this particular day we were

Pink and gold spectacular.

doing the rounds of the up-market clothing stores to find me a dramatic gown for the stage. I touched and tried on several thousand over-the-top outfits till we saw *the one* — we both knew it immediately. The skirt is made up of forty panels block-printed with gold paint, and it has a 28-centimetre, heavily painted border. Lining it is a stiff, multi-frilled petticoat to make sure it has lots of body. The first time I wore it on stage the audience stood up and clapped.

When visiting Jaipur, I hunted down the *varak*, the makers of silver and gold leaf, an amazing experience I wrote about in *Hot Pink Spice Saga*. *Varak* is probably from the Arab word *waraq*, meaning leaf or paper, similar to the Moroccan word for very fine pastry, *warka* — which I suspect has the same origin. The driver of the tuk-tuk I was in told me to listen out as we approached a certain quarter. Sure enough, there was a symphony of banging filling the streets of an area studded with old buildings, with open, basic rooms containing bare-chested men, sitting cross-legged and hammering what looked like small pieces of leather. Happy to be observed, these friendly craftsmen invited me into their workshops, which were filled with even more noise from their chatting, singing and transistor radios. I discovered that they were

What is a girl to
do when faced with
such luxury?

pounding small booklets made of dried bull intestines. Between each page of the booklets were little pieces of silver or gold being beaten to tissue-thinness, just a few micrometres thick.

The boss's office was on the third floor in his flat, where he lived in just two brightly coloured rooms with his wife and the younger of his five children. Bedding and clothing were neatly stacked on shelves. The kitchen was set up in one corner. In another room his wife sat cross-legged on the floor placing the impossibly thin leaves of silver and gold between cut-up phone-book pages to make booklets. Being so thin, they are extremely fragile, so even the smallest breeze or heat of fingers will cause the leaves to disintegrate.

Again food and fashion come together, for these precious leaves are sold all over the world for two distinct purposes: being edible they are used to decorate those exquisitely presented Indian sweets; but also there is a long tradition of them being printed onto fabric. *Varak* printing started in the fifth century BC and was mostly done for royalty (just as with elaborate dishes of food) and also for embellishing textiles to be used for sacred purposes.

Nowadays, this fabric is used for very special occasions, such as weddings, or for high fashion, but being so expensive there are very few practitioners. I went to a lecture in Delhi given by Dr Vandana Bhandari, a professor at the National Institute of Fashion Technology, New Delhi, and she said she was aware of only one person still producing it. The patterns of leaf are printed onto fabric using carved wooden blocks, perforated metal stencils or a fine brush. The fragility of the leaf means washing the textiles is virtually impossible, but they are no doubt very rarely worn. They become heritage items, passed down through generations.

Of course, I had to go in search of such clothes myself and found them in a shop in the City Palace called Satayam, owned by the Sharma family. What is a girl to do when faced with such luxury? After all, I had to ensure this rare art is continued, so there was no choice but to purchase a fine white cotton coat they had made. And I have managed

There's no need to restrict gold to jewellery: how fabulous is my white Andraab cotton coat printed with gold flowers?

to wash it successfully, I just had to be very gentle and not use an iron on the gold bits.

Khari is another old but less expensive technique that prints with gold and silver dust or, even cheaper, sparkling dusts like mica (a sheet silicate mineral) and chamki (rough ruby rock). The effect is not too far off the much more expensive *zardozi* metal-thread embroidery. However, the dust sits on top of the fabric rather than permeating it. To work with the dust, the stamping tools have two parts that fit into each other: the outer part is called the *sancha* and is a brass case with the pattern perforated onto it; the inner part is called the *hatha* and is a wooden mallet. Again, there are not many printers who continue this art, and most are based in Rajasthan.

Kalamkari block-printing

Another way to elevate fabric is through block-printing. Wooden blocks are also used in the more common technique of printing textiles with dye. Teak blocks are carved with gorgeous patterns, dipped in a beautiful dye hue and pressed onto the fabric, which is stretched out on long tables. All done by hand and by eye, this process is repeated

Not quite a Marilyn Monroe billowing skirt, but this blue Kalamkari block-print swirls out beneath my Brigitte Singh blouse with no need of an air-vent.

with every different colour and part of the design. The hand-blocking technique requires a lot of time as well as water because the first stage involves week-long soaking of the cloth to remove oils and any loom lubricants. Echoing *varak*-making, there is plenty of beating involved, too, this time to swell the fibres so they will readily absorb the dyes. Next the cloth is treated with *harda*, a thick yellow astringent paste made from the dried and powdered plums of the myrobalan tree, which acts as a pre-mordant or fixer on the cloth. After the dyeing is done and the patterns printed, the fabric is washed and beaten all over again.

Persian Mughals brought the technique of block-printing to India, and the East India Company later took it to Europe to become '*les Indiennes*', as outlined in the following section.

Rajasthan is yet again the place to buy hand-blocked fabric. Just outside Jaipur, in Amber, you can find Brigitte Singh's bougainvillea-covered workroom, *haveli*-style home and shop. She arrived there in 1980 from France to study Indian miniature painting, but fell in love with the master's son, who gave her the name Singh. Having been captivated by Indian chintzes, she decided to design textiles the way they used to be made in the seventeenth century. Her master block carvers and

A Brigitte Singh printed skirt, beautifully tucked and finished, with earrings echoing the flowers.

printers hand-produce high-quality cotton and *chanderi* voile, and she hand-finishes the clothes, accessories and furnishings with tiny stitching and buttons. Her refined and intricately made clothing, accessories and furnishings are sold across the globe in exclusive boutiques and are among the best in the world. And, being the best, it will be no surprise to you that I have been wearing her garments for years. Again, washing them is a potential drawback, but I don't wash them by hand in cold water as I'm supposed to, and, despite all sorts of mistreatment and plenty of wear by yours truly, they still look perfect.

A more middle-of-the-road shop in Rajasthan is Heritage Textile and Handicrafts, run by the wonderful Paliwal family. They do block-printing and will make garments to order, so of course I order lots of different light *kurtas* — a long tunic split at the sides and worn with pants. Men's clothing, such as shirts and *kurtas*, is just as popular and flamboyant. The prints are of flowers, birds, leaves and other flourishes. Another producer to look out for is Anokhi, who have reasonably priced shops all over India and a museum in Amber, Jaipur. Owner Faith came to India from America and, in another love story, fell for John Singh. They opened their first shop in Jaipur in 1984, and even back then were determined to be socially conscious, treating their craftspeople with honesty and transparency and using vegetable-based dyes. They also sell appliqué, embroidery, patchwork and beadwork.

On the culinary tours I host in India most years, we do a block-printing workshop at Bagru, about an hour outside Jaipur, with master dyer and printer Vijendra. Everyone block-prints their own scarf and takes it away with them, accessing their inner Picasso with joy. Vijendra also makes my scarves under my own label, and everything is done sustainably. The design was created by Bella Akroyd of phd3 in Auckland, then carved into blocks, dipped in non-toxic dyes and stamped onto silk/cotton-mix fabric. They are un-hemmed to show they are handmade and not factory produced. The procedure of getting my scarves printed in India was an exercise in extreme patience. The whole village has one job and that is dyeing and printing fabric. The place smells of dye, there are block-cutting shops everywhere, areas of bare earth the size of cricket pitches are covered in drying fabric and people are up to their knees in rinsing vats. My printer's wife makes very good chai. We sit on the floor and smile at each other.

I showed Vijendra the design of my scarves on my laptop, and he got a piece of paper and quickly drew it. Next time I turned up, the wooden block with my design had been carved. Looking at the colours on my laptop, Vijendra grabbed a few pots of pigment and mixed the two colours for the design. He dipped the block in the dyes, stamped an example on some silk/cotton cloth and said, 'Is this okay?'

'Yes,' I replied, 'that's perfect and very beautiful.'

Then he got his expert onto the printing. They carefully printed up a hundred or so scarves and that was the end of that.

'Aren't you going to write down the recipe for the colours?'

'Nope,' he smiled.

'How will you know what to do next time?'

'I'll remember.'

Showing the design on my laptop to having the scarves in my hands took over a year. When you buy a scarf you are buying a story.

Péro blouses combining checks and flowers, embroidery, contrasting linings, beautiful buttons and the trademark heart — what's not to love?

My absolute hands-down favourite Indian designer is Aneeth Arora of Péro. From Udaipur, she learnt her trade, as did many other Indian designers and textile makers, at the National Institute of Design in Ahmedabad and the National Institute of Fashion Technology in Mumbai. Her clothes are exquisitely beautiful, feminine, handmade and bursting with embroidery, flowers, floating layers, checks, craziness and, most of all, wearability: all are loose and comfortable.

The fabrics are created, printed, dyed and woven by her company. Everything is inspired by traditional Rajasthani dress, but she somehow makes it all very chic so you don't feel out of place wearing it in your home town. You know how usually when you buy clothing when you travel it looks fabulous in the country of origin but you just get pitying glances when you wear it on the Champs-Élysées; not so in Péro.

Les Indiennes

Another notable printing approach is found in *Indienne* fabrics, which were traditionally decorated with flowers. From around 1500, colourful fabrics were sent to France from India, where the stunning revelation that you could actually print on fabric, and in such bright and lasting colours, meant the French quickly became obsessed by these seemingly magical bolts of cloth. Previously you could have flowers on your dress only if they were hand-embroidered, so unsurprisingly most people lived their lives in plain outfits — but not any longer. Within a century, printed fabric from India, dubbed '*Indiennes*' or painted cloth, had become standard for such things as clothing material, household linens and upholstery. As shiploads of textiles arrived in France, shiploads of cash left for India, and French manufacturers went under; so much so that King Louis XIV banned the product before the country went bankrupt. However, the ban was eventually lifted and French manufacturers started producing their own versions of the printed fabrics themselves. They brought the blocks in from India and set up printing workshops, which operated for the next few centuries until cheaper mass-produced printed material took over in the early 1900s.

Thankfully businesses like Souleiado and Olivades have rekindled the art of hand-printing *Indiennes* with original wooden blocks from the old days. It's very time-consuming. The first step is to bleach the cloth, then heat and dry it. Designs first rendered on paper are stencilled onto the fabric through little holes. Colours are added by the carved wooden blocks and retouched by workers (who are usually women) called '*pinceauteuses*'. Finally, the fabric is washed.

Cotton

I visited Ahmedabad in Gujarat in 2019, first for food and second for textiles. It's a big old fifteenth-century city with fantastic Islamic architecture and heart-stoppingly gorgeous carved wooden *havelis* (mansions). I went through lots of them. They had double courtyards, several storeys, and are full of frescoes, indigo walls, painted floors, secret passages — they are the sort of houses you can see yourself restoring and living in. Mahatma Gandhi comes from Ahmedabad and so does the prime minister, Narendra Modi. I went to Gandhi's home by the Sabarmati River; it was very simple and spartan. I wasn't invited to Modi's place, but I did stay in the same hotel he stays in, the wonderful old The House of MG. All the money for these great houses comes from textiles. In the interests of research I spent hours in fabric shops. In India, fabric shopping is fabulous — you sit on the floor or a low platform and the salespeople spread everything out before you. You say '*khadi*' and there's a dramatic swish of cotton, you say 'linen with roses, please' and veils of fabric float down before you, you say 'traditional block-prints' and colour falls at your feet.

I also graced the Calico Museum of Textiles. The sixteenth-century word 'calico' comes from the city of Calicut (Kozhikode) in Kerala and refers to untreated, unbleached cotton. The museum is a very beautiful collection of *havelis*, and the exhibition of antique and contemporary cotton, fabrics, textiles, clothes and art is brilliantly curated. Unfortunately the staff were so rude, so unaccommodating and so entirely free of charm that I burst into laughter in the middle of the

We all know that New Zealand was built on the sheep's back (and its more delicious cuts).

Wool

The principal difference between cotton and wool is that wool is made from protein and lipids and cotton is mainly cellulose. Wool has a crispy texture that gives it more bulk than other textiles, and enables it to hold air and heat. Interestingly, it also absorbs sound, so if you sing to your jumper it will retain the tune. I made that up.

Being a New Zealander I am an expert in sheep jokes. These you will be spared, but I will talk about how much we love wool and how much the rest of the world loves our wool. We all know that New Zealand was built on the sheep's back (and its more delicious cuts). Currently we have 27.4 million of them to be precise (and only 4.8 million humans). They thrive because of our temperate climate, sufficient rainfall and extensive land suitable for pasture, meaning grain diets and nutritional supplements that guzzle fuel and leave residues are unknown to the local woollybacks. We are blessed with knowledgeable farmers who are well versed in the best grass types, farming techniques and genetic technologies to ensure our breeds deliver the goods. They are also increasingly employing sustainable agricultural practices so our natural resources of good land, air and water are preserved. The result is sheep — and therefore their products — that are healthy, nutrient-rich, lean and free of undesirable substances.

The wool industry produces about 1,160 million kilos a year — most coming from China, Australia and New Zealand — but it only comprises 1 per cent of the global textile market. Wool is actually a heroic fibre. Everyone bangs on about sheep belching methane, but wool accumulates carbon from the air, retains it and only lets it go when it has completely decomposed (which only takes a few years). It will even completely biodegrade in seawater. As it has a high nitrogen content it can be used as a fertiliser. Wool is a fabulous fibre to wear — due to its ability to absorb and release water just like hair, it will keep you warm in winter, cool in summer and it absorbs sweat then releases it. If you have been in a smoky pub all night, just hang your wool garment on the clothes line for an hour and it comes up smelling

The smocked Akira jacket, made from a wool and silk mix, this time being expertly modelled.

of roses. We wash our clothes far too often — most natural fibres will smell great after airing. My new travel rule is if it doesn't look dirty and doesn't smell dirty then it isn't dirty. I learned this airing trick from a cameraman when I was shooting my food show in a 40-degree Indian village in Rajasthan. Every day he wore the same clothes, every day he poured with sweat and every day the clothes were like new. Also in terms of sustainability, wool garments last *much* longer than any other fibre — think of all the jumpers and shawls you have had for years and years and are still wearing. Think of all the jumpers you knitted for boyfriends and husbands — they're probably in perfect condition in someone's attic.

You get wool off an animal by shearing it with electric shears. In ancient times it was just pulled off with the hands or a bronze comb. Hand shears were invented a long time ago — most likely as far back as the Iron Age. Once you've got the wool, it then has to be scoured or cleaned of the lanolin (sheep fat), dirt, sweat and general muck sticking to a sheep. It is then separated out into grades, the top being what's called 'fine', which usually comes from the shoulder. Really fine soft wool is used for clothing, while the coarser grade is made into coats and carpets. After washing comes carding, done by a machine where tooth rollers tease the fibres, open them up and produce something more even and workable. The fleece is then combed, spun into yarn, then woven on a loom to create fabric. What's new now with wool is that it is treated so that it doesn't shrink when washed. Organic wool is big, most of it coming from New Zealand — for example, Stella McCartney buys her wool from here.

Of course, wool doesn't only come from Merino sheep — New Zealand, for instance, uses other breeds like Romney and Drysdale. Cashmere and mohair come from goats; alpacas, llamas and camels also produce wool; and angora comes from rabbits. Fashion lines such as Lacoste and Monsoon have stopped selling angora products because of the cruelty involved in obtaining the fur. Alpaca wool, on the other hand, is mostly cruelty free as the animals benefit from having their fleece sheared, though it is all down to how the shearers treat them. When I searched for traditional alpaca goods on my visit to Bolivia in 2003, I saw alpacas on the menu in restaurants, though among the cheap copies of European style I did manage to find and buy a lovely knitted alpaca hat and gloves.

Alpaca wool became all the rage on the cat walk about five years ago. Labels, such as Ulla Johnson, Rochambeau and Nicholas K started featuring garments made from this super-soft luxury fibre. Meanwhile designer Michelle Peglau, who originally came from Peru, set up Hortensia Handmade in New York, selling clothing made from 100 per cent alpaca wool hand-knitted by Peruvian artisans. The wool has none of the scratchiness of sheep's wool, is flame resistant, hypoallergenic being lanolin free, lightweight, sustainably produced, and can be spun to be thin for summer or cosily thick for winter. Alpacas are much kinder on their environment than destructive cashmere goats. The most expensive type of wool available is from a vicuña, which also hales from the Andes, a wild beastie that is a relative to llamas rather than alpacas. They are all camelids, a family that also includes the guanaco, and they all spit when angry. The alpaca is the woolliest of the family and is the domesticated variant of the vicuña, which is the least woolly, producing less than half a kilo a year of the most incredibly soft fibre, which is why it is so expensive (up to US$3000 for that half kilo). The llama is the biggest of the camelids and a domesticated variant of the guanaco. Llamas do provide wool, but they are mostly used as pack-animals and plate-animals. Given how silky soft the wool is, I have the perfect excuse to quote some Ogden Nash:

The one-l lama,
He's a priest.
The two-l llama,
He's a beast.
And I will bet
A silk pajama
There isn't any
Three-l llama.

Linen

Flax springs from the earth which is immortal; it yields edible seeds, and supplies a plain and cleanly clothing, which does not oppress by the weight required for warmth. It is suitable for every season, and as they say, is least apt to breed lice.
—Plutarch (c.AD46–after 119), *On Isis and Osiris, translation 1936*

Linen, made from silvery-gold flax, is what the ancient Egyptians used to wrap their mummies and cult statues, leading them to consider it sacred, powerful and magical. Linen was connected to seclusion, secrecy and ceremonial protection. They not only bound mummies in metres and metres of it (up to 16 layers) but they removed the internal organs of the person, cleaned them, wrapped them also with fine linen and placed them in jars.

In ancient times, flax for linen (because it also produces linseeds and linseed oil) was planted in October and harvested in March while still slightly unripe. The harvested stems were dried in the fields for a

It is best just to give
in to the creasing
and call it a look.

few days before being 'rippled' — beaten and shaken to get rid of the seeds — then 'retted' or soaked in water to soften. Next they were dried, before being 'scutched', or beaten to separate the woody stems from the flax fibres. The process continued with 'hackling', which is when the fibres were combed to be split, straightened and cleaned, ready for spinning. Those are four verbs you never knew before.

The spinners would first roll the fibres between their hands and thighs, rolling it into balls, which they would then spin by holding a spindle in each hand and work with one leg raised so they could roll the spindle down the thighs to twist it. This is not unlike the Māori method of preparing native flax for use in piupiu and weaving. (Celebrating that fabric, the New Zealand Prime Minister Jacinda Ardern wore a flax cloak to Buckingham Palace in 2018.) All this thigh rolling was a very sensuous activity with close body contact, which not only transformed the flax but made the thighs calloused, a bit like a guitarist's fingertips. The weaving was then done on horizontal wooden looms, often producing very high-quality fabric with 50 x 30 threads to the centimetre. Egypt is famous to this day for the high thread-count of its linen and cotton sheets.

Linen is beautiful against the skin, breathes, feels clean and is very comfortable to wear. It is moth-resistant and wicks moisture away from the body. It reminds us of languid summer days, just lying around in crushed cream pants and sandals, while holding a mint julep. It is strong, lasts much longer than cotton or silk, softens with washing, but creases easily. It is best just to give in to the creasing and call it a look. It's highly biodegradable and sustainable as it takes a lot less water than other natural fibres to produce, and if it hasn't been dyed or bleached is 100 per cent biodegradable. In terms of fashion, linen goes in and out — one season it dances, the next season it's passé. It has a tendency to be taken up by unglamorous groups like straight, middle-aged ladies on cruises — not exactly edgy and frankly verging on frumpy. Recently, however, because of the whole sustainable issue, designers have been making an effort to freshen up the image: think well-cut, fitting (enough of the loose already) black top and skirt with black-and-white espadrilles, outrageous earrings and cat's-eye sunglasses. Linen/cotton mixes are popular, and I recently bought a linen/silk-mix fabric in India that is amazing — it looks fragile, feels like silk but is very strong.

Silk

I first saw silk being made in Hội An when I was writing my book on Vietnam, *Noodle Pillows*. Walking around the town you often hear a loud clackety-clack clackity-clack in the back of a shop. I walked into one and discovered an ancient world of cotton and silk weaving.

Silk-making was the closely guarded secret of the Chinese for centuries, ensuring them great wealth — and death to anyone who got too close to the source — but with the collapse of the Tang dynasty in 907 the mystical silk world was revealed to everyone. The Vietnamese then became the master silk weavers, in some places making even finer silk than the Chinese. So there, standing in front of me was a seventeenth-century wooden silk loom clacking away. Upstairs, baby white silkworms were squiggling around on a huge flat bamboo basket in a wire and wooden cabinet. They start their lives off as tiny black dots, sleep for three days, then turn white. They eat finely sliced mulberry leaves every three hours during the day and twice a night for three days, then rest for one day while they moult. This goes on for about 16 days till they become '*an roy*' or 'devouring' and are moved to a bigger cabinet with larger leaves, freshly replenished by the staff every day.

For eight days they gorge themselves in an orgy of gluttony. When they reach their adult size (the size of a small woman's little finger), they carry a full stomach of silk inside their now red bodies. The worms are then placed on fine branches enclosed in a bamboo rack so they can spin their silvery cocoons. They do this over four days by spitting out the silk thread and winding it round and round their exhausted bodies from the outside to the inside. Although the worms are only about 5cm long, each one can make about a thousand metre-long thread, 30 micrometres wide — half as wide as a human hair. They are making their protective home, which looks like a quail egg, in order to stay inside and turn into a chrysalis. In an ideal world, the chrysalis would turn into a white moth and burrow its way out, already full of unfertilised eggs. The females immediately mate with the males, then lay about 500 tiny eggs on a mulberry leaf, at which point the whole natural cycle would begin again. *However*, in the world of silk-making, the chrysalises are murdered inside their cocoons by being boiled in water. Next you just pick up a cocoon and pull a thread out from the base and it goes on and on and on. This impossibly fine, almost invisible thread is wrapped onto spools, then woven into fabric. It takes 20 kilos of cocoon to produce 1 kilo of raw, silvery white silk. When you see how it's made you can't believe it's not more expensive.

Like cotton, there are lots of different types of silk depending on the weave, including charmeuse, which is particularly shiny on one side and matt on the other side; gauzy, lightweight chiffon, created with twisted thread, which makes it akin to the heavier georgette and the thin organza; crêpe-de-chine with its more muted finish; taffeta, which has a rib pattern and can be reversed; crisp, strong, nubby-textured dupion, which is woven with two threads, and if the threads are different colours the dupion is called shot silk; and glossy silk satin, among others. Did you know that once velvet was made from silk? You can still buy silk velvet, but it is likely to be a blend of silk and rayon.

When you buy fine silk to have a dress made, you have to know how to choose it. These are the methods of divination:

- The old wives' tale that silk can be pulled through a ring is true, especially with scarves and smaller items.

- The cost is a giveaway — if it's too good a deal for all that work from the industrious worm, then it is likely to be fake.

- Tug the fabric, and if the threads stay tight it is the real deal; if it shows elasticity, the cloth could be chiffon but is more likely synthetic.

- You might not want to do this to a finished piece in a shop, but if you ask for a sample of the cloth and burn it, silk will smell natural, and its ash will be loose, while synthetic material will melt and result in lumpy ash.

- Silk threads are angular, so the dye won't be solid and will have a sheen that's slightly multi-coloured when you angle it to the light.

- Silk will crease a little if scrunched up, and even make a crunching sound, while other fabrics will either crease a lot or barely at all.

- Grab an iced drink before you go into a shop, so your hand is cold, or take a sample home and leave it in your freezer, then hold the cloth; if your hand gets warm quickly, it's silk. If you can't do that then just try rubbing it — silk should quickly become warm.

- Some fabrics may still be silk, but not from carefully farmed mulberry worms; for instance, tussah silk is from wild worms, producing less consistent, coarser thread; silk noil is nubby-textured and comes from the leftover short fibres from combing; while habotai silk is a lower grade of silk that has been sand-washed to give

Every year (barring Covid) I host a culinary tour to Vietnam because it is famous for its sophisticated, zingy cuisine, and, of course, I can't resist its sparkling, if rather conservative, fashion scene. I wrote about this in *Noodle Pillows*. When I first had some silk clothes made there, I learnt a lot about their way of operating — and I think it also taught them something about Western capriciousness. In those days, shop clothes were made for tiny Vietnamese women, not buxom European ones, so the likelihood of finding a good garment off the rack was as minuscule as the average Vietnamese waistline. Even though I was only size 10, I knew I was still 300 times bigger than any local woman. My solution was to have a garment made to measure. For some reason, all the guidebooks made this process sound easy: you just go in, point, and within hours have a fabulous, perfectly fitted three-piece suit or a copy of Marilyn Monroe's Happy Birthday Mr President dress. Nothing could have been further from my experience. It took hours, days, kilowatts of patience and a powerful fan to stop me going through the roof and orbiting a few planets. Maybe if I had ordered something standard, it might have been easy, but by now you'll know I am far from standard.

In my chosen Hanoi silk shop, the fabric was folded in piles on the shelves rather than on bolts, which is exactly as it is in India. So, after a long discussion about the outfit I wanted, there was another long discussion about which fabric, followed by another about the colour:

'I would like this lime-green silk for my Chinese top and pants, please.'

'Oh no, Madame, you are very beautiful but too old and too fat for this colour. This good colour.'

They show me a muddy green.

'No, thank you.' So I asked my friend. 'Do you think I look fat in this colour?'

'No. You look like a fluorescent light bulb, but not necessarily fat.'

'I would like piping on the pants, please.' The muddy green comes out. 'Oh no, I want red.'

'*Red*?' Giggle attack.

'Yes. Red as can be.'

Next, the staff drew a picture of my dream and measured me in humiliatingly loud detail in the middle of the shop. Finally, they said, 'Come back tomorrow at 5pm and all will be ready.' This, I came to understand, was one of the biggest falsehoods in the universe, after 'the cheque's in the mail' and 'but darling, it's you I really love'. I was leaving Hanoi the very next day, so when I turned up at 5pm, it should have been no surprise that the clothes were not ready and the ones that were didn't fit properly. All I could do was sit and wait, sipping tea under the fan. For several hours.

Because I'm a slow learner, I did this all over again in Hội An. The owner of the chic shop that I chose was encrusted with diamonds and was perfectly made-up. Her staff wore identical outfits, which I discovered changed daily. Again I was given tea in tiny cups, and this time told I was beautiful and asked where my diamonds came from. I realised they did tough deals with gentle smiles. I ordered a red Chinese jacket, a kimono and a black silk dress with fake-fur trim — but only after the owner had approved my choice. She ruthlessly assessed my body, skin, personality and diamonds to judge what would be best for me.

'No, no, no. You can't have that silk, it's not the right one for that style of dress — you need this one.'

'No. I want this hundred per cent silk.'

'No, believe me, this silk mix much better.'

'But I want this.'

'No. You take this.'

It might amaze you to hear that I did what I was told. Vietnamese are conventional, so I was on a hiding to nothing hoping they would step out of their square. They've always made garments this way, so this is how it will be.

'I want longer sleeves.'

'No. Look silly like that. This design must have short sleeves. I stake my career on it. I do this all my life, Peta, trust me . . . I know better.'

'Okay.' Needless to say, I never wore the dress because the sleeves were too short, but *c'est la vie*. What did I know? They also made me Chinese thong sandals of all hues, always with the dire warning: 'I can't sell these crazy colours to anyone else, Peta. If you don't like when finish, you pay.'

Silk is often the chosen fabric for the *ao dai*, the traditional dress of Vietnam. Along with the Indian sari, the *ao dai* must be one of the most beautiful national dresses ever invented. This sexy, elegant dress is half closed and half open, yin and yang. The skirt part is split from the waist on either side so it accentuates the breasts and the hips (yang) and hides the abdomen and legs (yin). Walking in an *ao dai* creates a flow of fabric, like a cloud floating along the street.

The *ao dai* was first worn in 1744 when it was decreed that both men and women should wear trousers and a gown that buttoned down the front. In the 1920s, the king modernised the design, and a decade later the artist Le Pho made it more body-hugging with a lower collar. Finally in the 1950s two Saigon tailors produced an *ao dai* with long, straight sleeves, creating a diagonal-buttoned seam from the collar to the underarm, the style everyone wears today. The pants are fitting at the top and flare at the bottom.

In the North in the 1960s, the *ao dai* was rejected as being inappropriate for hard revolutionary work, a ban lasting till the 1980s, when it began to be worn again for formal occasions, or as a kind of uniform in the hospitality business. When I visited in 2002, one of the defining images of street life in Ho Chi Minh City was the beautiful flying angels: girls dressed in *ao dai* and riding on Honda OMG

(Genio), accessorised with long gloves, mask, sunglasses, sometimes a conical hat (to keep their skin as white as possible) and high-heeled mules or platforms. Given the length of the skirt, *ao dai* are not the easiest to accommodate, but these girls had it down to a fine art: when riding pillion, they sat side-saddle; when riding singly, they sat on the back flap of the dress and daintily grasped the front flap in the left hand on the left handle of the Honda. Saigon girls wear them very long over vertiginously high platform shoes because they all want to be tall.

As with the sari, I had never really been tempted to wear an *ao dai* because I knew that if I wore it down the local in my home town I would just look ridiculous. UNTIL. Until I met Diego Cortizas and Laura Fontan, a Spanish couple who arrived in Hanoi in 2004. Like me when I arrived in Paris in 1980, they fell instantly in love with Vietnam and it changed their lives. They were enchanted and inspired by the colour, food, people, fabrics and crafts they found around them and decided to do a mad thing: start up a slow fashion company that reflected all this — a sort of Spanish/Vietnamese fusion they would call 'Chula', meaning a pretty, gorgeous, cute woman. Diego was an architect and artist, and Laura asked him to design clothes for her — that's how it started.

Diego is handsome and warm, and Laura is tall and beautiful with a magnetic personality. They now have three gorgeous children who are just like them. The kids grew up on the cutting-room floor amidst parties, openings and concerts. Chula is sustainable with zero waste, and they create furniture and accessories with left-over fabric. Their philosophy encompasses 'inclusion, cultural traditions, happiness of life, sense of humour, handmade fashion and suitable small-scale projects'. The person who buys a Chula garment is buying not fashion, but a work of art that is so well made it lasts a very long time.

The main workshop is in Hanoi, in a big house with a shop and café, and is chock-a-block full of sewers, cutters, stylists, tailors and finishers. And it is silent. There are 65 employees and 80 per cent are deaf or handicapped in some way. Laura and Diego have learned sign language, and everyone is madly chatting and smiling all day, but there is no sound so it's a very peaceful place. Laura says the staff's sewing skills are extraordinary — they can do anything.

The clothes are intricately embroidered, appliquéd, block-printed and beribboned. The designs are wild and crazy, and walking into a Chula shop (they're all over Vietnam and have pop-ups and fashion shows across the world) is like walking into a Picasso painting on fiesta day. We go there on my Vietnamese tour, and the minute my group walks into the shop they are gone — I have lost them, they can't see or hear me. They dive on the clothes, the men buy shirts, the women buy party dresses, and I see them wearing these clothes for years on social media. They have dinner parties back home where they wear their Chulas and cook the Vietnamese recipes they have learned. We go upstairs and meet the workshop sewers, learn how to make Diego-style block-printed scarves and, because they have absolutely nothing else to do, Diego and Laura cook us a *paella* for lunch.

One day in 2017, I walked into the shop on West Lake, fanning myself and looking for a cup of tea. I was stopped in my tracks by the most fabulous *ao dai* in the front window. It was white silk taffeta with silk organza sleeves over wide black silk pants. It was painted by hand all over with an intricate mosaic design in rusty colours and blue featuring a Roman goddess. I asked if I could try it on.

Diego said, 'Oh no, it's just a display. I made one only for a fashion show we did in Rome.'

I looked at him. He smiled at me. The girls measured me up, set to hand painting a copy, and had it delivered to my hotel by the time I got to Saigon. That's what you call service. I have worn this stunning dress all over the world and still wear it to the last-night bash of my Vietnam tour every year.

Ordinarily, I live half the year in a very beautiful medieval town in the South of France (Occitania, or what used to be Languedoc, to be precise) called Uzès, in the *département* of the Gard. Near us we have the lovely Cévennes mountains and the whole area has lots of mulberry trees. These huge trees not only produce delicious berries but also leaves for silkworms to feed on. In 1266 the beautiful secret of silk finally reached the Mediterranean, and by the end of the thirteenth century there was the plantation of millions of chestnut and mulberry trees and the beginning of silk production in the Cévennes. Just like the story of Indian block-prints, which the French were so in love with, they were also importing enough silk from India and China to break the French government's budget. So just like the block-printing, the French decided to do it for themselves. Thousands of families were involved, and almost every farm and vineyard also produced silk, with a large magnanery, or silk workshop, added onto the farmhouse.

Production in France peaked in 1853 with 26,000 tons of cocoons, which was woven into 5,000 tons of silk. Several thousand Protestant communes, providing more than 300,000 workers, were involved in the industry. Over time the silk industry disappeared completely due to disease in the worms, wars, political instability, and finally the death blow when synthetic silk arrived in France. In 1965 the last mill in Saint-Jean-du-Gard closed down and that was the end of that. Today China and India are the biggest producers of silk.

Incidentally, you can also get silk made from different species of silkworm, such as eri or muga silk, or from different types of creatures altogether, such as spider silk and sea silk. Spider thread is very strong, extremely fine, golden in colour, and exquisite expensive garments have been made from it. Sea silk was made by early Egyptians, Greeks and Romans and also in Arabia and China, where it was known as mermaid silk. It is made from filaments secreted by large molluscs to attach themselves to rocks or the sea-bed. It is even finer than ordinary silk and yet very warm. As there is only one sea-silk weaver known to be still alive (in Sardinia), a garment made from sea silk is unlikely to be turning up in my shed anytime soon. There's also a vegan silk made in Morocco from aloe vera found in the Sahara. It is highly prized and called *sabra*.

Fur

The history of humans wearing animal skin and fur is very, very ancient. Forty thousand years ago, Cro-Magnon man started wearing the fur look, partially for decoration and partially for warmth. Modesty didn't come into it, but as time went by status did. Even Ötzi — the stone-age Italian man found in 1991 preserved in ice — was wearing wonderful patchwork fur pants, loincloth, coat, shoes and hat, and he lived sometime between 3400 and 3100BC. The way he is dressed is not primitive — it's stylish and funky, sewn together with bone needles, so he was obviously expressing himself creatively. He had 61 tattoos, mostly in the form of lines and dots, possibly as guides for acupuncture. His clothes and shoes were expertly made and designed, so researchers think there were cobblers who made shoes for other people.

But back to status. Eventually the most sought-after skins and furs — like leopard, lion and ermine — came to be available only to royalty and the ruling class, and lower beings were forbidden to wear them. There was also the early concept of 'contagious magic', where the wearer of furs from certain animals would hope to assimilate the qualities of that animal, for example, bravery, robustness, skilfulness, etc. So how did fur move from the upper classes to the general population? People who could afford fur wanted to have the right to wear it for decoration, status and warmth because there is nothing like a fur coat in freezing temperatures to keep you cosy, even if are just a boring old commoner. Originally only wild animals were hunted for fur and pelts, but in the past century, as the demand grew, animals (called furbearers) began to be farmed specifically for their skins. There are still, however, wild fur harvesters, and one of them is in New Zealand — but more later.

The most famous high-fashion fur in the world is from the Italian house of Fendi, operating since 1925. Karl Lagerfeld was their chief designer for 52 years. They still produce the most stupendous, expensive, fabulous fur coats, dresses, shoes, hats and stoles, so outrageous they're practically unwearable — think brightly dyed with the addition of feathers, beaded flowers and pompoms. If all this seems élitist and self-indulgent, gob-smackingly beautiful as these garments are, they are mostly for inspiration because what people really buy is a trickled-down version that doesn't actually cost a million dollars. My paternal grandmother, who was wealthy and travelled to Europe for her holidays, had a black fox stole, a white mink jacket and various fur coats. I remember her getting dressed for balls and formal occasions. I can almost smell that stole — during the summer it was stored in mothballs and tissue paper in a cool cupboard downstairs. If you choose not to wear fur because of concern for animal rights, don't bother going for fake fur — it's not biodegradable.

Haute-couture fur coats are sourced from such unlucky creatures as fox, mink, sable and chinchilla, but let's turn to bunnies. Rabbits were introduced in the 1830s to New Zealand's beautiful, pristine environment (the only mammals here before humans arrived were bats, or those in the sea). Rabbits were shipped in to provide food, fur and sport. This bright idea resulted in rabbit plagues and caused an ecological disaster, with damage to native fauna and flora. Within 20 years, great swathes of the South Island's high country became dry, bare and barren from the impact of bunny grazing and burrowing. The problem is that rabbits breed like rabbits to the tune of 45 kits a season, meaning populations proliferate about tenfold per annum.

For 100 years there was actually an industry for the meat and fur (the fur being particularly fashionable), and 'rabbiter' was a real job and export production a thing. The industry was de-commercialised in the 1950s, and methods to control rabbits haven't been very successful ever since, often getting bogged down in tussles between animal welfare and agricultural viability. A number of farmers cull the rampant rabbits through sharp-shooting, and the hand-selected skins used in my friend Jane Avery's eco-fur coat business, Lapin, are sourced from them. This results in beautiful clothing and fewer destructive bunnies.

Jane worked with Mooneys Furriers in Dunedin to make her garments. Mooneys began in 1912, and the last owners were Max Wilson and Heather Kirk (they have just retired). Max learned from

Stunning one-off coats
from Jane Avery made from
repurposed embroidered
vintage sari silk from Rajasthan
along with New Zealand wild
rabbit eco-fur, with silk lining,
and wool and cotton interlining
(© Jane Avery).

his furrier father in Manchester and had been in the trade for 50 years. The rabbiters who shoot the rabbits for Jane in the South Island high country approve of the fashion option because the rabbits have to be killed anyway so you may as well use them. From this land of extremes came the motivation for Jane to transform an environmental pest into fashionable, warm and luxurious clothing.

Along with personal hands-on service at her studios in Dunedin and Central Otago, Jane's bespoke business encompasses long-distance commissions with the sending of drawings, samples and individually tailored calico mock-ups to ensure customer satisfaction before commencing a make. The coats, which can take up to six weeks to manufacture, can use any number of rabbit skins to produce. Jane also uses gorgeous vintage Indian textiles along with the fur. Her latest project is 'ReVintaging' where she restyles client's old fur garments to make them fashionable, or even changes them into something completely different, like a cushion or a bed throw. Fantastic idea. Have a look at her website: https://www.lapin.nz/.

Jane: 'Mooneys Furriers were just that: furriers. They made coats and blankets and cushions and hats and other furry stuff, buying in furs when they needed them for commissions or accepting contracts from tourist-related fur retailers. They didn't deal in skins. That's tannery business. Furriers buy in furs from tanneries. Or in the case of the big wide world of fur, from wholesalers and auctions such as Saga Furs in Copenhagen. I buy my furs from a rabbiter who lives around the corner from our property in Alexandra. He shoots on the big high-country stations. Also from a guy with a pet-food factory near Christchurch. He shoots rabbits up in the Mackenzie country. They hand select the best furs they can for me from their cull and send them to a tannery in Invercargill. Then they're sent to me ready to be worked with. There's quite a bit of handling involved — which is one reason why it's expensive.

'Max and Heather, the furriers, were so generous in mentoring me and allowing me to use their equipment, such as the pneumatic stapler and boards for stretching the skins and a cup seamer, which is a specialised fur-sewing machine. When we first met over five years ago, Max prepared and sewed my skins, and Heather stitched the fur into my first couple of coats. But we soon realised that wasn't economic with their time or my money. So they adopted me and I became their apprentice, if you will. I've inherited their furrier equipment. So kind. I love them to bits. Now they've retired, I'm continuing the furrier tradition with rabbit and possum and vintage fur, too. This can include remodelling a client's vintage furs and incorporating the fur with my own design work. Max, thankfully, continues to mentor me. I must continue learning from his expertise while I can.

'I have no intention of purchasing furs that aren't introduced New Zealand pest species but am happy to do remodels and restoration of people's vintage coats. Working with long-dead fur from bygone eras fits comfortably with my world view. I refuse to engage with the anti-fur lobby as I think their views are inflexible and lack nuance and context. I guess when I began Lapin a few years ago I had a more fixed viewpoint about which animals it's okay to exploit for fur. Introduced species' pest fur is obviously my thing. At the other end of the scale, animals farmed in cages, such as minks and foxes, make me feel uncomfortable. Wildlife management fur such as coyote in North America, I think I'm okay with now even though they are an indigenous species. I try to read a lot and be informed on the complexity of the fur debate because I find it interesting philosophically. As a topic it impassions so many people, but there's also lots of hypocrisy and ignorance.'

As touched on, another animal plague in New Zealand is the Australian brush-tail possum, introduced 150 years ago for the fur industry. Unfortunately this furball eats millions of tons of native leaves a year, so something had to be done. Same story as the rabbit. Possum fur is actually rather wonderful to wear if you can get the word 'roadkill' out of your mind. The fibre traps a cone of air, which makes it lighter (14 per cent) and warmer (8 per cent) than, say, wool. Polar bear fur has the same structure. Possum is often mixed with Merino lambs' wool, which is very soft and light and feels like cashmere.

CHAPTER 8

Shoes

Before shoes there were feet. Walking is something you don't need to learn how to do — a baby just walks when the time is right and its body figures out how to balance things naturally. The moment we learn to walk is the last time in our lives we actually walk properly, and that is because of shoes. Shoes change how we walk. It took 4 million years for our human foot to reach its unique composition, and by association our way of walking. When we walked barefoot, our feet were much healthier — to this day Zulus, who often go barefoot, have the best feet and the most correct walk in the world. In 1960 Ethiopian runner Abebe Bikila won the Olympic marathon in Rome in bare feet.

Our feet propel us forward through life, and mostly we are wearing shoes on them. Chances are, you are wearing some sort of shoe right now. Maybe you are reading this book on a beach wearing flip-flops, maybe you are at home in soft slippers. You can't wear slippers outside — they are poofy and cosseting, not meant for hard surfaces. Maybe you are waiting for a bus wearing strong boots to protect you from the weather. But shoes are so much more than practical coverings for the feet. One of the best master shoemakers who ever existed — Roger Vivier — said: 'To be carried by shoes, [to be] winged by them. To wear dreams on one's feet is to begin to give reality to one's dreams.'

I could create a biography around the shoes that have marked my life. Even now my grandnieces' baby shoes create the same emotional response as did my niece's baby shoes, which I commented on a number of books ago. As I've also written about previously, the first pair of shoes I fell in love with and wanted to marry were my ballet pumps when I was five. The ballet shoe is a flat, soft shoe made of satin or leather, specifically for dancing. Romeo said to Juliet: 'You have dancing shoes with nimble soles. I have a soul of lead so stakes me to the ground I cannot move.' I understand that soul of lead when I remember having to relinquish my pink satin ballet pumps: they evoked the smell of the lacquer and the joy and elation I found in dancing. I loved their softness, their fragility, the pink satin ribbons attached to them, the tiny little bow at the front, the stiff midsole pressing snugly along the bottom of my foot. When Mum painted them to refresh the colour, the perfume was like a happiness drug (I think there was alcohol in the paint).

For a lot of important events in my life, I can likewise remember what shoes I was wearing at the time:

- Strapped at school: heavy, rain-soaked lace-ups (in black and nothing else).

- Nursing: heavy, rain-soaked lace-ups (in white and nothing else).

- Debutante ball: silver lace-up pumps (my 'dancing shoes with nimble soles').

- Fell in love for first time: sandals with straps to the knee (the Roman slave look was in, okay?).

- Wore new boots to bed in Canada: cowboy boots from Seattle (let's pass on quickly).

- Got married: black patent-leather shoes with red stiletto heels (there had to be red).

- Gastronomic tour of Northern Italy: Pucci mules (Emilio Pucci was the Marquis of Barsento but preferred to be a fashion designer).

- Fiftieth birthday party: 15cm Perspex platforms (light but sweaty).

- Seventieth birthday party: sparkly kitten-heel mermaid shoes (what else, darling?).

In my book *Can We Help It If We're Fabulous?* I wrote that 'scientific studies have shown that if a woman has no lust for shoes she is probably dead from the neck down'. It was then that I began my exploration of the allure of footwear and found out just how complicated it is to hand-make a good shoe.

Here's a rundown for any budding cobblers out there. The 'last' is made first. You might have seen one in a junk shop; it's a handmade wooden replica of a foot, made for each shoe style. The last prescribes the height and contour of the arch and where the body's weight will be distributed across the foot. Like a perfectly whipped mousse, a well-designed shoe won't just taste and look good, but will feel good. We've probably all experienced how miserable we're made if our shoes are too tight or too loose, if they squash toes or rub heels. You know a good designer when you're not hobbling down the road in your brand-new shoes. So, to avoid being a sadist the designer needs to consider the fit as well as the look of various parts of a shoe:

- the toe box (an obvious label)

- the vamp (a not-so-obvious label for the area above the toes)

- the throat (a label that couldn't suit my food analogy any better)

- the heel breast (anatomically confusing, this refers to the area on the inside of the heel)

- the shank (again anatomically confusing, this refers not to the leg but the arch support).

Got all that? The designer hands over to the pattern-maker, who cuts out the shoe's upper and lining and creates the toe box and the counter (which stiffens the back of the shoe). The leather is soaked to be malleable enough to stretch over the last to be nailed in place. Like a Christmas cake doused in brandy, the leather is left for a couple of weeks, before the sole and heel are attached.

The next stage is, of course, the dangerous one: when those beautiful, shiny, sweet-smelling creations are placed on a well-lit glass shelf in a high-end shop with a come-hither ambiance. Who can resist, even if you're an average woman who, according to a UK 2013 study, already has 20 pairs back home? Apparently the not-so-average women can own a hundred pairs or more, carrying on Imelda Marcos's legacy.

In an article in the *Journal of Research in Personality*, Omri Gillath, Angela J. Bahns, Fiona Ge and Christian S. Crandall wrote about their investigation into whether you can judge a person purely by looking at photographs of their shoes. They found that their subjects 'accurately judged the age, gender, income, and attachment anxiety of shoe owners based solely on the pictures', and so they concluded that: 'Shoes can indeed be used to evaluate others, at least in some domains.' My mother didn't do any research, but she always used to say you could tell a lot about a person by the *state* of their shoes.

Shoes have, throughout the centuries, been symbols of wealth and status, but my mother was really saying: never date a man with shoddy, dirty shoes because he will be a shoddy, dirty person in general. If only I'd remembered that. Little did I know that men have a completely different outlook on shoes in general — they are not in love with them as we are, and don't seem to use shopping for them as a form of therapy. They don't sit around fantasising about shoes all day, and they don't think a fabulous shoe will change their lives. They don't own uncomfortable ones just because they are gorgeous, and they don't seem to own 3000 pairs. Interestingly, knock-out women's shoes are completely wasted on men — they couldn't care less about your Jimmy Choos or your Robert Clergeries. They wouldn't even notice that Christian Louboutins all have red soles. Cheap shoes are fine with them, and they wouldn't know the difference. Possibly excepting the Beatle boot in the 1960s, I can't imagine a male shoe becoming fetishised the world over in the way the Manolo Blahniks and Jimmy Choos Carrie wore in *Sex and the City* were. No.

I had a French boyfriend when I lived in Paris who did, in fact, care about his footwear. Alain was a chef (normally the worst-dressed people in the world) who wore the most elegant, expensive shoes in the kitchen, on the street, at the beach, in a nightclub. I even remember the name — J.M. Weston, French master shoemaker since 1891. His preferred style was the Oxford shoe with the medallion cap toe, contrasting stitching and traditional perforations on the front and sides. I never saw this boyfriend wear anything else. He said they reminded him of boarding school and made him feel secure. He always wore white socks, white pants and navy cashmere sweaters — he had dozens of them. Why I didn't think this was a clue to his mental state, I don't know. He was horrified by my casual hippy way of dressing (it was the early 1980s) and completely transformed my look, teaching me how to dress like a Parisian.

Life is too short to
wear boring shoes.

Kartell shoes from a Milanese
furniture manufacturer.

My French publisher's office was in a sartorially risky part of Paris, the 2nd *arrondissement* — full of Jean Paul Gaultier, Kenzo, Stephane Kélian (shoes), Robert Clergerie (shoes) and Maud Frizon (shoes). In fact Maud Frizon made the best shoes in Paris: beautifully constructed, very quirky and long lasting. Now in her eighties, she has shops all over the world. I gave my sister a pair in 1985 and they are still avant-garde.

Which reminds me of my Kartell rubber and plastic sandals. In 2017 I was in the fabulous, glamorous, incongruously named Bon Marché department store in Paris when I came across these glossy plastic sandals. The friend who was with me said they were the most ridiculous things she had ever seen. What do friends know? I had to ask every shopper in the store if I was doing the right thing, then decided that life is too short to wear boring shoes. I love them because they bring together fashion, art and design. They're coquettish, eclectic and ironic. Paula Cademartori created them for Kartell (who normally make furniture), made from plastic injection moulding they are very clever. The hands that are joined to form a heart are a reference to the theme of universal love, and the flowers were inspired by Rio de Janeiro's botanical gardens.

High heels

I am five foot four, so high heels can make me the required six foot. They also force your centre of gravity forward, arching your lower back, protruding your buttocks, lengthening your legs, accentuating the shape of your calves and raising your arches. This may sound akin to being on a rack but, according to sex researcher Alfred Kinsey, it is actually identical to female orgasm. Furthermore, in the 'courtship strut' of the animal world (no, they don't wear heels), erect ankles and extended legs are the signal for sexual availability.

So, high heels are sexy and seemingly even help exercise your pelvic floor muscles. But then, as is always the case when there's good news about one's vices, there's also bad news. Apparently, our bodies are not designed for high heels, and we could end up with all sorts of toe, foot and back deformities from wearing them, not to mention squashed internal organs from our tipping pelvises, endangering our chances of conceiving.

Being well trained by both the convent and hospital sisters, I nod sensibly, note my passing years . . . and promptly purchase another pair of high heels. Okay, I do reserve the really high ones for only the occasional special occasion, and have a lot of low ones in my collection, but I will not banish high heels from my life altogether. After all, they make you feel fantastic, especially well-constructed, designer ones. There is an enormous difference between cheap copies and the real thing. Like our feet, designer shoes are made individually, so matching them up to get the best fit can take time. When individually made, the heel size differs according to the size of the shoe, so with a smaller shoe, the heel will be proportionally smaller. With cheap shoes, the heels are generally all one standard size, meaning a smaller person will be trying to balance on a shoe with heels made for someone larger.

Which brings me to my mermaid shoes, which I bought on the Paris tour with Aloïs Guinut. These foxy mermaid shoes are the most whimsical, delicious pair of shoes ever invented, made by Italian designer Alberto Gozzi. They are described as rainbow sandals, but the chic

High heels by Marc Jacobs, who in 2010 was one of *Time* magazine's 100 most influential people in the world and in 2012 was one of *Out* magazine's fifty most powerful gay men and women in America, which shows the power of fashion.

Aloïs immediately dubbed them mermaid shoes and it has stuck. The front, pointed part of the shoe is decorated with shimmering greenish sequins; the back part is open and attached onto the foot with a black and shocking pink strap with a buckle. The kitten heels are wooden. As I wandered all over Paris in the freezing cold, my sensible shoes started seriously hurting, so I put on the mermaids and was very comfortable for the next few hours — they are soft like slippers and it's like wearing nothing. Interestingly these shoes look best with straight-leg pants.

Not everyone will want to walk miles in heels, but there is a trick to it. Just like mastering chopsticks when all you've ever used is knives and forks, you have to adjust more than expected. Don't walk as you would normally in flat shoes. Your body should follow the shape of the shoe, rather than making the shoe bend to the way you walk. Centre your weight on the heel and take smaller, slower steps. Think of deportment classes, even if you've never taken them: head up, shoulders back and look straight ahead. Even still, you might not be able to bop till dawn in high heels. Celebs and models buy shoes a size too big for them to allow for the fact their feet will swell as an evening progresses. Gel pads are inserted to stop them from slipping off.

My mermaid shoes
by Alberto Gozzi.

Don't wear high heels on grass, but do wear them with slim pants and jeans. High heels immediately formalise whatever look you're sporting. If you've had to go straight from work to a function or dinner, just putting heels on and redoing your lipstick shows that you mean business. I'm very into kitten heels — the shorter stiletto curving in from the back of the foot — because they've got that *je ne sais quoi* chic and you can also walk easily in them.

Needle-heeled stilettos first appeared in the early 50s, created by Salvatore Ferragamo and Roger Vivier. Needless to say, Marilyn Monroe was quick to wear these sexy creations. She is reported to have said: 'I don't know who invented high heels, but all women owe him a lot.' Stilettos have stabbed their way in and out of fashion over the subsequent decades. As have platform shoes, which were worn in ancient times, were popular in the fifteenth century (useful for lifting you above the muck of the street), were ubiquitous in the 1970s and have come and gone ever since.

Salvatore Ferragamo also invented the wedge. This was a heel from an earlier era than the stiletto — first appearing in 1936 — and championed by an earlier bombshell: Brazilian Carmen Miranda. She was only five foot tall, but you'd never know given her high exuberant headdresses and glittering 20 centimetre wedges (she knew how to make fashion work for her). A wedge is obviously a wedge shape, its heel filled in with a light material like cork, providing a safer and firmer centre of gravity.

These days a lot of espadrilles have a wedge, too. I loooove espadrilles. Espadrilles mean summer, *pastis*, happy-peasant-at-one-with-the-people, easy living and brown legs. Being made out of canvas and string, they are both disposable (because you throw them away at the end of summer) and chic (being redolent of sunshine, bullfights and *tapenade*). They are perfect for walking across cobbled streets (which is impossible in other heels). Originally called *alpargatas*, Spanish farmers made the soles from esparto grass, before they caught on in the south of France and became espadrilles. When I was in Morocco, I found myself inexplicably sucked into the shoe shops. There they will make you shoes by hand, including a green, beige and red pair made from raffia — a different take on the espadrille.

Boudoir shoes

On the one hand, the slipper may evoke Cinderella and Marilyn Monroe — on the other, Hilda Ogden. Originally designed for wearing only in the bedroom, they were sensual, delicate and finely decorated, made from silk, satin or velvet (a far cry from the modern-day sheepskin Ugg boot).

Slippers might make a public appearance these days at weddings in the form of soft, flat shoes similar to ballerinas' slippers. My favourite pair are the ones my friend Sue gave me: soft black suede lined with soft white wool and topped with big bunny-coloured pompoms. She turned up at a Christmas party with armfuls of these slippers, so we all padded around in them for the whole evening.

The Japanese have a whole range of slippers, from jandal-types with wooden platform soles to ones made from paper. Meanwhile, in North Africa they have the slipper-like *baboushes*, which come in all kinds of colours, some with pompoms on, others with beads, or in the case of my pink and yellow pair, a tassel that could be shaped as you pleased. The leather is so soft, but can be worn outside, and some people might call them mules.

Mules were originally the ultimate boudoir shoes: high-heeled with a naked back foot and a feathered strap across the toe. The go-to feather for not just mules but for trimming dresses, dressing gowns and hats from the nineteenth century on came from the down under the marabou stork's tail. These feathers were soft, floaty and easily dyed (often in pink). In reality other birds, such as turkeys, also provided the goods. The coquettish Diana Dors was often depicted tottering around in feathered mules. Since her day, the mule has also stepped outdoors and the feathers taken flight in favour of embroidery, pattern or plain leather. Even the heel has become optional.

Charles & Keith mules, redecorated by yours truly.

Recently I was in Singapore airport and bought a pair of Charles & Keith mules because they had pointy toes, which I always think are so elegant. But I hated their brown colour and rest-home-pink bow, so I took them home, had them dyed turquoise and attached my own southern French bow.

The slipper worn by footmen in the sixteenth century gave rise to the court shoe (originally worn to the royal court) or pump (from the French word *poumpe*, worn when visiting the pump room to take spa water). Women quickly realised they would make dancing a lot easier, and before long were wearing the pump outside and in. Today the court shoe reveals more of the foot and is attached to a stiletto in bright, sexy colours, or else is a different breed altogether, being low-heeled, usually black, favoured by business women, Michelle Obama and those attending funerals.

A word here on flat shoes: they look good with cropped pants, skirts of any length and volume, pantyhose, tights, shorts. Chunky flat shoes like ankle boots, platform sneakers, loafers and shoes with straps make your ankles and legs look heavier, so are best worn by people with slim ankles and legs, unless you're wearing them with long pants. Delicate shoes like ballet flats, pointy shoes and sandals always make your ankles look finer than they necessarily are.

Kate Spade mules. This inspired designer died tragically young in 2018.

- Under no circumstance should you wear white shoes, ever; they do nothing for your style, are a nightmare to keep clean and make you look like a nurse.

- Don't let holes develop — get them fixed when they start to wear thin, and resole as soon as it's necessary.

- Wet footwear should dry naturally, not in front of a heater, which will damage the leather and could cause it to separate from the base.

- Alternate the shoes you wear day to day to give each pair time to breathe and dry out.

- Clean stains with vinegar and water; scuffs can be removed with a rubber.

- If your boot zip gets stuck, run an ordinary pencil along the zip.

- Refresh and reinvent old laces with different colours and fabrics. If your laced shoes are causing discomfort, try re-lacing them in a different pattern (for instance, straight across instead of diagonally, or vice versa); the pressure on your feet might well be miraculously relieved.

- Take your boots off when you go to bed, even if you are commitment-phobic. Leaving them on is only for cowboys.

An example of *kanthar kaaj* on a single layer of tussar silk (a style local to West Bengal).

CHAPTER 9

Accessories and Garnishes

The iconic designer and couturier Hardy Amies maintained that budget-strained people should always wear expensive accessories, as they reflect glory on cheap clothes. If you are of a conservative bent in your dressing style, accessories are a safe way of doing something dramatic without walking into a party all guns blazing in gold lamé. This is where your shocking-pink stilettos, lime-green bracelets and leopard-print pillbox hat come in. We will suffer sore earlobes for fabulous earrings, we hide things in coat pockets to be forgotten and found years later, we wear a shawl rather than a cardigan to feel graceful, we flick open a fan at the slightest hint of heat. It is through accessories that we betray our secret dreams, inward landscapes and memories (I used my grandmother's handbag for years because it made me feel close to her). Our bags hold the most personal things, without which we think our lives will come to a standstill. One day at a party, a woman's husband was so furious with her, he just opened her handbag, tipped it upside down and let the contents fall to the floor. I was so shocked that he had exposed such intimacy in front of everyone that to me it was worse than if he had hit her.

Evening bags

Evening bags are a good excuse to add a *soupçon* of glamour, and you don't have to confine them to the evening. They can be beautiful, fun and opulent. For over the top, you can't go beyond designer Judith Leiber's jewel-encrusted evening bags, which come in a whole array of shapes, such as soda cans, cell phones, lips — there's even a teapot. Meanwhile, Alexander McQueen designed a beaded clutch whose handle is made up of four jewelled rings, so you can wear your bag. In a further elevation of the humble bag, artists have recently been employed by designer brands to design clutches that are not just a work of art but highly desirable collector pieces.

Many of my evening bags are Indian, all loosely fashioned in the traditional *potli* style, which is a small drawstring purse. This style is ancient in Indian culture, going back to the Vedic period (c.1500–c.500BC). The Hindu god of wealth, Kubera, is always pictured in ancient scriptures and paintings carrying these bags (usually full of money or beads), and Lord Rama's brother Lakshmana filled his with fruit and medicinal leaves. When the Muslims invaded India around the fourteenth century, they brought with them a sort of *potli* or saddle bag in which they carried samosas for travelling by horseback. The hero Arjuna carried magical bows and arrows in his. These days the purses are mostly embellished fashion items, usually made in silk, velvet or satin, decorated with embroidery, fringes and tassels, and inlaid with lots of bling, such as foil-work, sequins, mirrors, beads, stones, and — for those with money to spend on the outside as well as to place inside — pearls and diamonds.

My evening bags might not carry much inside but they sure carry a lot of style.

One of my favourites is pink silk, lined with red and white dots and tied with silver tasselled drawstrings. I bought it in Udaipur, in the Leela Palace Hotel. Then I bought a pale-blue and gold block-printed one lined with pink silk in Jaipur, at Rasa. Another I use a lot is embroidered soft leather from Morocco. But my absolute favourite bag is the most gorgeous chartreuse silk satin embossed with hundreds of tiny steel beads, which I bought at a luxury dress and jewellery shop called Bombaim in Kolkata.

Shawls

I never bought or wore shawls before I went to India, but once there I was enchanted by the graceful and nonchalant way Indian women wear them. I wrote a piece on shawls in my book *Hot Pink Spice Saga* as they are so much a part of their everyday lives, even in the field, desert, jungle, work gang, kitchen, factory or palace. They are worn for modesty, warmth, decoration, symbolism and to carry things like babies. A lot of Indian women wear the three-piece *shalwar kameez*, which is probably Arab in origin. This is composed of pants (*shalwar*), a mid-calf length dress or tunic (*kameez*) like a *kurta*, and a long scarf (*dupatta*).

Some of the Indian beauties that seduced me into becoming a shawl wearer.

The word 'shawl' comes from the Persian word *shal*. They were worn by both men and women, as they still are today in India, Afghanistan, Egypt and Iran. These *shal* were always a luxury item. By the eighteenth century, they had been brought into England and France, where upper-class women went mad for them; they wore them over their voluminous dresses to very beautiful effect. The soft Kashmiri shawls were so popular that the Indian weavers and embroiderers couldn't keep up. The Europeans then decided to copy them and produce them more cheaply, faster and with slightly inferior wool.

One of the most well-known motifs is the one called paisley, probably depicting a bent cypress tree or *boteh* and symbolising prosperity and fertility. It is an ancient teardrop design from Persia, and when the British started copying it around 1820, they called it paisley after the town of Paisley in Scotland where the shawls were made.

My *dourukha* (which is Persian and means 'having two faces') cashmere shawl comes from a shop called Andraab in Jaipur and is double woven by hand by two weavers working simultaneously (meaning there is no back side — both sides are perfect). These shawls can take years to create — talk about slow fashion. The material is called

An Andraab cashmere shawl (right) next to a *Les Indiennes* print.

shahtoosh, is known as the 'king of fine wools' and is made from the under-fleece of the Tibetan antelope or *chiru*. It is so fine that it can be easily pulled through a small finger ring. It is light but very warm — if you sleep under it, it feels like you are sleeping under flower petals. This is an heirloom shawl meant to last for life, then be handed down to family — the designers advise not to wash it unless you really have to.

Andraab cashmere shawls are made in Kashmir, India, an ancient snow-topped land with ruined temples and legendary Mughal gardens. They have sumptuous cuisine too; it is very rich and perfumed. The shawls are made from the finest grade of *pashmina* (the name comes from the Persian for wool). It is hand-spun by women into gossamer threads from the hand-combed hairs of Changra goats, native to the Changthang region, northern India and surrounding areas. The hand-weaving techniques came into India from Persia in the fourteenth century (along with block-printing, elaborate cooking, jewellery-making and Muslim culture). The weaving is normally done by men. Real cashmere is a rare, luxury product and very expensive — when you buy a garment you are buying a work of art. Andraab was started by the Andrabi brothers in 1999, and they focus on restoring the lost magnificence of pashmina. The shawls might also be intricately embroidered with super-fine silk thread and gilt, silver and gold wire, requiring thousands of stitches. All the dyes they use are natural: the yellow from marigolds; the red from pomegranate skins; blue from indigo; orange from saffron; kohl for purple; verdigris (from copper) for green; but a lot of it is left plain in its natural colours of ivory and browny-grey, the colours of the goat. What makes these shawls special is the harmony, brilliance and depth of the colours. There are peacock patterns, Mughal florals, minute patterns, zigzags, stripes.

I have another cashmere shawl I bought from Kashmir Loom in Delhi, which is embroidered. The work is done in the villages around Srinagar in Kashmir. The technique is part of the Kashmiri Sufi tradition, where embroidery becomes a form of meditation; focused, meticulous attention in which every tiny stitch is part of a jewel-like pattern. The first step in embroidering is to carve a raised design on blocks of walnut wood just like you would for block-printing. You dip the block in a rice/gum paste then mark it onto the shawl. At the end of the procedure, these marks wash out. The fine silk thread comes from the worms of the local mulberry trees, and there are various embroidery techniques used, like *kanikar* and *soznikar*. When it is all done, in the most incredibly tiny needlework, the shawl is washed in mountain river water, dried, then steam pressed, ready for your tender shoulders.

My silk scarf from Rasa in Jaipur is block-printed with blue dots, gold painted on top and filled in with embroidery.

Kantha

Kantha work is done in the eastern regions of India, particularly West Bengal. I had long been enchanted by this embroidery, so was very interested when my friend Husna-Tara Prakash introduced me to Shamlu Dudeja of SHE Foundation (a self-help enterprise) in Kolkata. Shamlu is a very gorgeous, statuesque older lady who was a mathematics professor. Her husband died and she became seriously ill, all of which forced her to retire and threw her into the depths of despair.

However, she became interested in *kantha* embroidery and began visiting friends in the *kantha* villages to fill the void in her life. She empathised with the women's anxieties, anguish and everyday deceptions, and talking to them made her realise she could stop focusing on her own sadness and help other people. She decided to organise and hone the women's fabulous *kantha* work, give it a bigger profile and make it a really worthwhile enterprise.

After years of hard work, Shamlu raised a simple quilting stitch to the upper echelons of society. She helped to unshackle *kantha* work from its humble traditional village boundaries and put it on the bodies and walls of the great, mighty and stylish. Her daughter Malika has set up Malika's Kantha Collection (MKC), a marketing organisation to promote *kantha* globally.

I turned up at Shamlu's shop in her beautiful home with its park-like garden to find her dressed in a fully embroidered *kantha* sari. For years I brought my gastronomads from my tours to this hallowed sanctuary, and Shamlu would even have tea and biscuits laid out for us in her stylish living room. Just thinking of that shop full of the most beautiful *kantha* saris, shawls and murals makes me gasp — each piece of work is a labour of love and passion, weaving in its patterns the stories of the lives of simple women, their joys, their sorrows and their aspirations.

Kantha is the quintessential recycling story. It started centuries ago when Buddhist priests, or *bhikshus*, begging for alms were often given old fabrics that they would layer and stitch together with a simple running stitch, making a warm shawl or blanket for the cold months. Village women did this too, making blankets, cushions and quilts out of old cotton saris. The women's work was so fine and the stitches so tiny that air was trapped between the layers and the ruched effect became a thing of great beauty. There is an artistic, very sophisticated version called '*do rukha*', where you can't tell the reverse side from the obverse.

The origins of *kantha* are very humble, and even today village women sit together stitching and bitching, sewing their own designs, colours and stories into the embroidery. Sometimes a work is passed from grandmother to mother to daughter to be continued. Sometimes the work is very simple, with just lines of running stitch, and sometimes very complex and ornate, with the embroidery telling a story with flowers, birds, deities, trees and myths. The work itself includes a meditative aspect where the sewers take pride in attaining wisdom and understanding of their inner selves. The resurgence of *kantha*, first championed by the famous Bengali Nobel laureate poet Rabindranath Tagore in the early 1900s, has resulted in the growth of the concept of self-worth and *shakti*, the empowerment of women.

Of course there are stunning shawls from all around the world, and in all sorts of fabric from chunky tweed to gauzy chiffon, but even just a plain one can make a statement if it's in a contrasting colour to your outfit. There are multiple ways to tie them; the internet is full of suggestions, some very elegant, some just plain weird.

Scarves

Even more so than shawls, a scarf will draw attention to your beautiful face. Scarves were particularly popular a few years ago, a 2015 BBC report even declaring that they were a must-have accessory for powerful women: 'Wearing a big, draped scarf sends a message that you're comfortable, approachable and warm,' said brand and image consultant Mirella Zanatta. 'Wearing the right scarf in the right way will make you appear to be a refined person who has put thought into your outfit.'

One of my favourite garments is the dusky pink and gold Indian skirt I mentioned earlier. This wonder comes with a jacket and a long scarf or *dupatta*. Indian women feel immodest if they are not wearing all three pieces. It's not a bad idea to always wear a scarf as you can cover your face, carry babies, tie up firewood and slap men with it if they get too frisky. I don't often wear it with its traditional jacket and scarf, but reckon the skirt would look really good with a tight black polo-neck jumper and chandelier earrings.

Parisian women have this way of wearing a scarf like no one else in the world. The scarf is normally silk (light and fine), large and is

Another favourite from my scarf collection, this one is by Fragonard.

wrapped around the neck loosely until there's none left. The ends are tied in a chic little knot to the side. You practically have to stand on your head to get it right.

Here are some ways to tie a scarf.

> – Square scarf: I never know what to do with square scarves without making myself look like a peasant girl at a Romanian folk festival. According to my Parisian fashion guide Aloïs, you fold it in half into a triangle. Hold the scarf in front of you with one point facing down. Cross the other two ends around the back of your neck and bring them around to hang in the front. That's it.
>
> – Fine scarf: Hold the scarf up by one corner so it's long and pointy, then wrap it around your neck. Keep wrapping it gently and loosely till the ends disappear, and wear it poofy and loose.
>
> – The Loop: This is the one I wear the most. Fold your scarf in half lengthwise, put it around your neck then pass the loose ends through the folded end. Tighten so it fits snugly around your neck.

Coats, jackets, suits, cardigans, jumpers

All these covering garments give a sense of safety and comfort to the wearer. If you're cold or shocked and someone puts a coat or jacket around your shoulders, you feel cared for, protected from the outside world, hidden from harm. The person who 'cloaks' you also feels they are doing more than keeping you warm: they are sharing love and concern, as if the fibres could do more than just comfort you — they could contribute to your happiness. It's like wearing a hug.

Because I try to avoid winter, I don't wear coats, jackets and suits very much, but I do adore cardigans. For years Marilyn Sainty at Scotties made these fantastic tight little cardies from what was basically stocking fabric — stretch polyester — and dyed them fabulous colours. I bought one every year and still have an orange one, a blue one and a jade-green one. Sometimes they were two-tone and you could reverse them. The most enchanting part was the row of closely spaced buttons from the neck to the hem.

During the Covid-19 lockdown, I was stuck in a New Zealand winter instead of being where I normally am, in a very hot Southern French summer. I had no winter clothes, so I went into my garden shed and found a plastic box full of jumpers. I hadn't worn them for years and some of them dated from my 1980s life in Paris. Because they were such good quality, they all looked like new, were still fashionable and there was not a shoulder pad in sight. It was wonderful hitting the high spots of my lockdown neighbourhood in a Jean Paul Gaultier hooded sweater, summer pants with pantyhose underneath and powder-blue Doc Martens I also found in the shed.

Although not technically outerwear, suits have a coat-like aesthetic. I have worn very few suits, save for school, nursing and chef uniforms, but certain women seem to like wearing them to work, some with pants, some with skirts. I have, though, kept one fabulous couture chartreuse-green linen suit from my Paris days, which now seems impossibly tiny. When I had my restaurant in Paris, I lost so much weight from worry that I had a body like an anorexic adolescent, which is how I fitted into this chic suit. The label is Clementine Couture, and the outfit also had mid-thigh-length shorts that went with it.

In 1966 Yves Saint Laurent helped women steal the suit from men (who stole it from the army uniform), with the Le Smoking suit. My objection to suits is their relentless conformity, respectability and unadventurousness, but that seems to be what people like about them — a second skin with pockets. A suit is just a uniform, but some women

The Clementine Couture suit with shoes and earrings; the fact that they all match may seem dated now, but it's a look that comes and goes in fashion and can be tweaked by selecting different shades of the same colour, or even deliberately contrasting colours. Let your accessories have a conversation with each other and your outfit.

are turned on by that — any uniform will do really, even if it's only the postman.

Are men turned on by women in suits? The answer to that is: men are turned on by anything. A woman in a suit is someone who has taken the power from men but she adds emotion — she's seductive and authoritarian at the same time, she's cool and urbane. Women don't want to be like men when they wear a suit, they want to be better — they want to take it over and add flavour.

In the case of suits, jackets and coats — if it looks right, it is right. A perfect fit is a combination of good cut, good look and good feel. Never wear a short-sleeved anything — wear long sleeves and roll them up if necessary — this also applies to shirts. A short-sleeved shirt is just a horror, and frankly shows weak moral fibre. Does Rihanna wear short-sleeved shirts? No. Does Diane Keaton wear short-sleeved shirts? No. Does Brigitte Macron wear short-sleeved shirts? No.

Jackets are light but coats are heavy, and that's how you feel when you wear them. Coats are relatively uninteresting till you put them on: they need a body to give them personality. I suppose jackets and blazers are okay if you're of a conservative bent, and they can be exciting teamed

Like a sprinkling of garnish, these hand-crocheted earrings from Morocco bring a touch of colour.

up with a flamboyant skirt or dress. They symbolise calm, moderation and untouchability, but can also cover up a wild, duplicitous spirit — throw a jacket on and who knows who you really are. A tailored jacket over a plunging neckline dress gives so many mixed messages I'm dizzy just talking about it. One obvious message is 'I'm wild and crazy but don't cross me.'

Although I'd not count myself among their number, I reckon girls who wear jackets and blazers exude self-assurance with a good sense of humour. Structurally, these clothes can give you confidence you are not feeling (especially if they have padded shoulders), sort of forcing self-discipline on you, enabling you to be in control in a world that is out of control and covering up lumps and bumps you don't wish to be seen. Jackets iron you out and reduce your femininity, thus minimising your distracting womanly attributes. By buttoning up a jacket we are buttoning up our bodies and feelings and fears. By slinging a jacket over our shoulders we are suggesting abandon and rule-breaking.

You do need to find the right cut and length (see the section on body shape on page 67). However, bomber jackets are another matter — they look great on everyone — sexy and reckless.

Earrings

The obvious message jewellery conveys is one of wealth, but women are far more articulate than that. The Queen purportedly sends veiled messages through her choice of brooches (for instance, wearing a moss agate floral brooch that the Obamas gave her when she first met Trump). Former US Secretary of State Madeleine Albright wrote a book on her choices of jewellery. Although Lady Hale claimed she had no ulterior motif in wearing her spider brooch when she judged Boris Johnson's prorogation of parliament 'void and of no effect', speculation was rife and spider accessories became all the rage in the UK. The late Associate Justice of the US Supreme Court Ruth Bader Ginsberg indicated such things as whether hers was a dissenting voice or the verdict was a majority decision through the wonderful collars she wore over her black judges gowns. Michelle Obama made the news when she wore a necklace spelling out VOTE during the 2020 elections.

Jewellery choices can pack a powerful punch. They can also direct the eye: a bracelet will highlight your delicate wrists. Jewellery can complete an outfit, adding a vital splash of colour, bling sparkle or metallic sheen. It can do a lot of work, so employ whichever items appeal to you. My weakness is earrings.

When I took my Scotties Marilyn Sainty green velvet coat out of the wardrobe to write about it, I was looking for the label and put my hand in one of the pockets. Within were some very valuable ornate earrings I had lost years ago. They were Italian and I bought them because Michelle Obama did. The only reason my house is so tidy is because I went into every nook and cranny to find these earrings. This is another good reason to keep your designer clothes: if I had sold it I would have lost the earrings forever. This inimitable coat was loaned to me by Scotties Boutique for my *Taste New Zealand* show in 1996. It is chartreuse silk velvet, lined with a gorgeous silk damask. One day I wore it to film a TV cooking demonstration *without an apron* in an interviewee's kitchen. The next day the director said, 'Is that coat supposed to have a yellow line across it?' It turned out the kitchen owner had cleaned the kitchen

My earring mandala:
slipping on a pair of these
is such an easy way to add
beauty to every day.

bench with bleach; I leaned up against it and the coat was ruined. What to do? TVNZ bought it. I took the hem up, thus eliminating the stain, and I still wear it occasionally. As I rarely have any winters, its outings are few and far between, which is why it took me years to come across the earrings.

During our Covid virus lockdown, I was asked to contribute a mandala to a fundraising project for hospice and Starship Children's Hospital. It's focus was finding beauty in the everyday. At the time I was writing about my Indian earrings, so I made a mandala out of them.

Of course in India there is no such thing as too much bling, and I always encourage my gastronomads to go mad wearing exotic Indian jewellery when they're there as they will never wear it when they get home. When you wear Indian earrings you are wearing history and culture. Originally Indians wore jewellery because it was surety — you could always sell it if times got tight. For 2000 years, all the precious stones in the world came from India, notably diamonds, and they still trade big in diamonds, especially in Surat in Gujarat. Interestingly, India invented the diamond drill, a technique they taught the Romans. The work was extremely sophisticated and intricate and became even more so when the Mughals arrived in the fourteenth century. There are many stories told by the British of outrageous, unbelievably valuable jewellery dripping off wealthy and royal Indians — gold, pearls, rare blue diamonds, emeralds, rubies, sapphires. The costume earrings I love wearing are derivatives of this historical art.

Most Indian jewellery to this day is still made by hand. The styles of such jewellery echo the ancient geometric motifs and patterns that you'll see in Indian architecture, be it Mughal, Islamic, Byzantine, Vedic or any of the multiple other sources that make up the richness of the country. I mean, really, humans will hang any form of decoration they can find on any part of the body they fancy — ears, nose, fingers, genitals, toes, ankles. I like wearing dramatic earrings because it is a simple way of adding something very beautiful, romantic and bohemian to your look.

During lockdown, I made my friends put on earrings and lipstick for our Zoom parties. If you have a long neck you can wear long earrings, and if you have a short neck you can wear smaller geometric-shaped earrings, which sharpen the features. There is an earring for every face, and no earring has ever made the wearer feel dowdy or joyless — you are never too poor or too depressed to buy another pair of earrings. An earring is like a star in the sky — even the most unostentatious person will permit at least a gold stud. You should store jewellery in separate boxes or soft pouches so they don't damage each other.

Hats

I recommend the French beret, for it gives the impression of just the right soft toughness, a veritable wave of sophisticated brain matter. It is the kind of hat that inspires a person to grow into it, to become the person they never knew they could be.
— Meia Geddes, *Love Letters to the World*, Poetose Press, 2016

I have lots of hats because they're beautiful, but I don't wear any of them. I find them too annoying, too easy to leave behind, too likely to

Here's a uniform I'd be happy to adopt: traditional hats for Provençal lavender pickers, not so sure about the work, though . . .

mess up my hairdo. However, they will save your delicate complexion from the sun, they can make a statement, provide distraction on a bad-hair day and are great accessories for a photoshoot.

I bought my traditional lavender-picker's hat in Nîmes in a shop specialising in bullfighting outfits, hats, homewares and '*les Indiennes*' in the form of voluminous skirts, which are commonly worn with tight black jackets and high boots to ceremonies and weddings. My strong hat has a small crown, very large brim, black velvet chin ribbon and is finely woven with golden straw. It is nothing like the cheap sun hats found in markets and is my pride and joy, along with my pink-and-brown striped espadrilles.

Different hat styles suit different face shapes, which is the perfect excuse to try on every single one in the shop. You don't want to be swamped and lose your neck — a hat is there to enhance your best features (check what it does to your cheekbones and eyes). Don't just view the hat on your head alone, but observe how it works with the proportions of your whole body. You can use a hat to disguise your less desirable aspects, for instance it can increase your height if you choose a high-crowned hat, or shrink your appearance if you go for a wide brim.

An upturned brim will lift you, a downturned one will make you look small.

A hat is more than just a cherry on the top. Pairing a hat with your outfit is important — perhaps accentuating a colour or pattern. It can also set the tone: a smart hat will dress up casual clothes, while a beanie or baseball cap will dress down formal attire.

Earrings are best avoided or kept simple when you wear a hat as they just get tangled in the fabric and can make your look overcomplicated. You may find you have to wear your hair in certain ways to accommodate your hat and to avoid helmet hair. For short cuts, wear your hat further back on the head to show you aren't bald (though if you are bald, the hat is your best friend — let it do all the talking).

You'll recall I mentioned living in Willy Maywald's apartment in Paris. In 2013, there was a retrospective of his work at the Atelier-Musée du Chapeau, a homage to hats, showcasing his photographs of stunning models in a wide range of hats from 1936 to 1968. A catalogue was produced, which you can find being flicked through on YouTube: *Hommage aux Chapeaux*. If looking at that doesn't get you into a hat, then you're a lost cause.

Fans

I am a huge fan of fans — they are beautiful, elegant and practical, and they really work in cooling you down. I have dozens of them, buy them for my tour guests, use them for decoration, use them to hide from people and give them as presents. I spend a lot of time in hot countries and have delicious memories of travelling in Spanish trains where all the women were gracefully fanning themselves. In my home in France I have them lying around on all three floors so I never find myself without.

Fans date back to the sixth century AD, and through history have been used for all sorts of things other than cooling down: ceremonial purposes, association with mythical and historic stories, a method of secret communication, an accompaniment to dancing and performance, symbols of wealth and status, protection from the sun and fire (particularly important if you're wearing meltable make-up), keeping the flies off food. It was the Japanese who invented the folding fan — I love the swishing sound when I flick it open. They can be plain, painted with flowers and designs, covered in feathers and jewels, have pompoms and tassels hanging from them, inlaid with gold and silver. They can be made from silk, papier-mâché, ivory, wood, leather and bamboo.

CHAPTER 10

Shopping and Preserving

Are you one of those people who opens your wardrobe full of clothes and gasps 'I have nothing to wear?' One of the reasons you can't see anything in your wardrobe is that you really can't see anything — it's an over-crowded, mixed-up mess. If you go into a shop over-crowded with stock, you're less likely to see a thing, while a carefully curated store with fewer items lets you appreciate each item and is more likely to make a sale. It's the same with your wardrobe at home. The first thing you have to do is separate everything out into categories. As you are doing this, you will naturally come across clothes you want to get rid of. This is called culling. If you have problems with culling, call a therapist and have them be there as a support person. If you can't afford a therapist, use alcohol to bring yourself round —

there's nothing like a stiff gin to help you let go of clothes, inhibitions and grief.

You know how Marie Kondo says if you haven't worn it for a year, get rid of it? If it's very valuable or you are really emotionally attached, I say, keep it *but wear it*. Wear any outfit you're doubtful about for one day, then make your decision. If it's obviously dated (like it has huge shoulder pads) but you still love it, get it remodelled — there's nothing worse than walking around in an outfit where people say, 'Oh, 1986 must've been your happiest year and you've never got over it.' My memories are full of the beautiful clothes I sold or gave away and regret doing so. Just be ordered about it — know what you've got and where it is.

A lot of clothes sit unworn in your wardrobe because you never learned basic skills like sewing and mending. You can blame your mother and a deprived childhood for that. This is a slightly tricky one because it's quite expensive to get someone else to sew a button back on for you or reinforce a seam that is coming apart. If you have a sewing machine, it is so easily fixed and suddenly you're wearing that pretty Marni dress again. Learning how to repair your own clothes is an essential tool of fashion sustainability. Some designers (like Kate Sylvester) now offer complimentary mending.

If there are clothes and shoes that don't fit, don't suit or look dated, look at them carefully, clean them and sell, or give them away. Occasionally you have to admit that the war is over and you have to actually throw a few things out. Once you've passed on all unwanted stuff, make sure to keep the clothes you want in their categories — that way they are always easy to find and don't disappear into a dark vortex or get eaten by the wardrobe wolf. Indian goes back into Indian, pants go back into pants, etc. Use good-quality hangers and have a hanger for each garment (anything to avoid ironing). A bedroom floor is not a wardrobe. However, a shed can be, so if your fabulous, beautifully fitting garment is too obviously out of date just now, and you happen to have the space, you can store it away in an insect- and rodent-proof container to resurrect when it becomes chic once again.

Learning how to repair your own clothes is an essential tool of fashion sustainability.

Now get fast and loose with your looks — who says you can't put a Pleats Please dress over Pleats Please pants, topped off with a tight-fitting Hermès sweater and geometric Marni earrings? Who says you can't wear those pink lace-up shoes with jeans and a pink Marni top? What about that cross-your-heart stretch yoga top with a big gathered skirt and boots? The first thing I do in the morning is get dressed. I can change outfits four times before I find the right one for how I feel each day. I might walk around in one for a few hours, feel out of sync, and as soon as I change it feel balanced and right again. It could be a colour problem, a mood problem, a weather problem. Sometimes I just keep putting more things on till I look like a bag lady, then start reducing till I feel I can face the world. The second thing I do is thank God for my more-or-less good health and more-or-less tolerable looks. It does seriously amaze me how much better a person looks with full make-up on, but I mostly content myself with the minimum that won't frighten the little children on the street.

As I keep banging on: making a statement is what it's all about. I'm not talking about you making the same statement I would make; this is about making your own distinctive mark. And remember we're allowed to have different taste. As far as I am concerned, a look isn't complete without the vital finishing touches. For instance, I wouldn't leave the house without at least lipstick on. A great hairdo, a line of eyebrow, a rondelle of blusher, manicured nails and polished shoes transform you from dowdy to sparkling.

How to shop

Shopping for clothes is the same as shopping for food. We all eat food and we all wear clothes. If you are a gourmet *and* a fashionista, you may have observed that you can no longer be these two things unconsciously — you now have to think about the impact on the planet and your body; you now have to be aware of the origin of the products. You wouldn't put rubbish (fast food) in your mouth, so you shouldn't put rubbish (fast fashion) on your back. It is a documented fact that supermarket food throws itself into your trolley, so that when you get to the checkout there are all these mysterious items in there like truffle crisps (when you know there is a no-crisps rule in your home), shortbread (when you know you already look like a shortbread) and cheap sherry (when you know how disgusting cheap sherry is). I have a similar problem with shoes when I travel. When I get to my destination and open my suitcase, I find that my shoes have had sex with each other and multiplied.

The secret to good supermarket shopping is to (a) have a list and stick to it and (b) don't go when you're hungry. If you go food shopping straight after a meal, nothing will seem appealing to you. You will go home, realise your cupboards and fridge are full of food, and you will make beautiful meals for a whole week from what you already have. Same with clothes. Frock shopping is much easier and more fruitful if you have a plan and stick to it. If you don't need it, don't buy it unless you have a plan to pass it on.

With clothes shopping, many factors come into play:

- budget

- the weather

- your mood

- whether you actually need anything or not

- how far removed from reality you are about your body

- how much time you have available

- location — are you down the local mall or are you in the jungle in Goa?

Shopping questions to ask yourself:

- Does this garment sort of fall in with what I already have so I can mix and match?

- Do I already have something like it? Think hard.

- Does it fit me perfectly?

- Is it so glamorous that really and truly, there won't be many occasions on which I can wear it?

- What do I know about the designer — are they eco-friendly? There is such a thing as 'greenwashing', where generic eco-friendly words are used to describe the garment. It's a have in my opinion and the same as 'healthwashing', where poor-quality food is marketed as being healthier than it really is.

- Why is it so expensive? Check the finishing, look carefully at how it falls on your body. If it isn't well cut, it won't fall properly (the problem isn't your body).

- Why is it so cheap? What woman on the other side of the world is suffering in some god-forsaken slum factory to make it?

Buying a new dress or shoes is like having sex for the first time with a new lover — there's nothing to compare with that first thrill of something new, something happy-making, something that raises the self-esteem.

Designer jewellery, like these Marni earrings, can also be squirrelled away until you fall back in love with them again: they'll still pack a stylish punch decades later.

If you are a compulsive shopper, at some point you will have to use self-control and just refrain, and really it's just a decision. Once you have made the decision, you are released from the compulsion. To buy pointlessly is just a shame because it spoils the joy of occasional shopping. Just the same with lovers.

A lot of people shop online and get constant email reminders of sales from shops. These reminders are the work of the devil, so unsubscribe from them. If you're on a diet, you don't have food in your house that will tempt you — if it's not there you can't eat it. Same with online subscriptions. It's an incredible waste of time reading all those emails in your inbox — time that could be spent on meditation, yoga and bringing about world peace.

Fast fashion is the antithesis of style because you're just wearing rubbish with no cohesion or durability. One of the reasons I can still wear clothes I bought 30 years ago is because they are not overtly trendy. Buy clothes that suit you; *don't follow trends*. If tent-like dresses are the fashion, that doesn't mean you will look good in them. Those dresses actually look really good on tall, slim women. If you are voluptuous or shapely, you should wear well-cut clothes that hug your figure. If you have no waist, don't wear belts. If you understand what suits you, you will make fewer buying mistakes. Nevertheless, sometimes you can break the rules. Take the case of the yellow dress on page 145. As a small person I probably shouldn't be buying a full, baggy dress, but the dress is so beautiful that I believe I get away with it.

If you're in doubt about your look, ask someone really merciless like a gay nephew or similar. Normally you don't even have to ask — the look on their face tells you all you need to know. A mirror was invented some years ago that can not only say that you are the fairest of them all, but it lets your friends do the same thing. Invented by a designer in New York, it is a three-sided mirror implanted with special technology, which sends a video feed of you wearing the clothing you're dithering about buying to any digital device you like. As you twist and turn before this mirror (mercifully placed outside the fitting room), your mother, sister, best friend or significant other (who may be on the other side of the world or just in the café down the road) can view you in that exorbitant rag and bring you to your senses with a text that says: 'You have got to be kidding me'.

Maybe that's why these mirrors haven't spread across the world since I first wrote about them over a decade ago — they don't actually result in sales. And yet, they have some advantages. You don't have to ask your mother to wait while you wrestle into another outfit in the tiny changing room; oh no, instead this special mirror has touch-screen capability. By the mere press of a finger on the left-hand panel, you can instantly change your reflected costume to another. Meanwhile, the right-hand panel will help you select matching shoes or accessories. How can that be wrong?

When I go shopping it's very directed and I am decisive — I know what I want and I know it when I see it. Shopping is not a chore for me, and I can get it done quite quickly. I would never dream of going to a mall — waste of time and too much choice. I go straight to small shops I like or to designer shops that have a recycle section. I also try to buy at sales when I'm travelling all over the world. Occasionally, I'll buy something at full price, which is really stupid, but it happens.

The reason people buy expensive clothes at full price is either because money is no object or because they have to have the latest new thing (which fortunately they soon get sick of and pass on to someone like me). The question I always ask myself is: is this dress or pair of shoes absolutely drop-dead fabulous or not? If it's not gob-smacking, I keep walking.

There is enormous personal power to be had when you remove the word 'need'. If you can release yourself from the object of your desire, it gives you great bargaining power. Same with buying a house — if you can step back, there will always be another one. Same with buying an antique bowl at the flea market; walk away and you will probably end up with it at a better price.

A great way to shop sustainably is to buy local designers, whatever country you live in. For example, if you live in New Zealand and buy New Zealand you are buying top quality, providing jobs, profiting from transparent work practices (as we have strict labour laws), tiny travel miles and reduced use of fossil fuels. Be aware, though, that the fact of the matter is, because it's so expensive to make clothes in New Zealand, most brands have at least a small amount of their production either made or finished in China or India.

Some local designers have joined hands to do things together, as with Mindful Fashion, set up by Kate Sylvester and Emily Miller-Sharma from Ruby. It is an industry collective with many members, promoting 'long-term sustainable growth through responsible business practice' in New Zealand. They want a fashion industry that looks after the environment and prefers people over growth and profit. They are basically into shining a light, looking for solutions, holding people to account and educating. And all this with a sense of humour, a lack of guilt-tripping and a caring attitude.

I don't buy clothes online in general because I have to see if the fit is good and the garment actually suits me, but I know that a *lot* of people do buy this way, so here are a few places I've found. The good ones will accept returns if you don't like the garment. How do you know if the garment you are buying online will actually fit you? Most online sites have a guide with international sizing (it is extremely unhelpful that every country has a different sizing system), and bust, waist and hip measurements that go with the sizing chart. Trade Me, Ebay and Instagram are now HUGE — most young people buy this way for both new and recycled clothes. Of course, clothing companies and shops have had an online presence for a long time, but this has now been added to by people putting their own clothes up for sale. Good On You is a sustainable fashion website — you can sign up for their weekly newsletter and also download their app to get their ethical brand ratings. Also check out:

— The RealReal

— FARFETCH

— Vestiaire Collective

— Good Trade

— Thread Up

— Poshmark

— Rebag

— The Vintage Club

— Swap.com

— Depop

The joy of Miyake Pleats Please garments, but also of flowing fabrics like this Indian skirt, is that they can be rolled to be stored, making them perfect for travelling or putting away in your shed.

How to clean and care for your natural fabrics the sustainable way

If you look after your good clothes they will last longer. If you get sick of the sight of them, they will then be in good condition to pass on or turn into something useful, like a mop. Greenpeace has told us that a whopping 96 per cent of discarded clothing could continue to be worn or recycled or used for something else — all we have to do is look after them. When you've worn, re-worn and fallen out of love with your clothes, avoid the landfill and sell them to reduce your fashion footprint. To be green is to be stylish. Selling a denim Yves Saint Laurent jacket, for example, could save 400 litres of water and 1.89 kilos of carbon. Meanwhile, look after what you have, and don't use synthetic chemical-laden soaps or detergents to clean — try to buy organic detergents or make cleaning products. Baking soda, salt, white vinegar and lemon juice are your friends for cleaning clothes, household and cookware. In the days of my Paris restaurant in the 1980s, we used copper pots and pans and always cleaned them to shiny with a paste made from vinegar, salt and flour.

You know that little swatch of fabric you sometimes receive with

a spare button when you purchase a new garment? This is to test your washing detergent on the material so you can ensure it won't be damaged.

If you want to make your own detergent, mix together 14 cups of hot water, a cup of baking soda, half a cup of salt and a cup of castile liquid soap (castile soap is made with vegetable oil not animal fat). Shake it all up and rock your very own biodegradable detergent — half a cup per washing machine load. If you've got a cotton load you want to whiten (like sheets), add a cup of lemon juice or vinegar instead of bleach. If possible, dry clothes in fresh air rather than a dryer. These days I only wash clothes if they smell or are obviously dirty. Stella McCartney told me not to overwash. The average washing machine uses over 50,000 litres of water a year — you'd need to drink for your entire life to consume that amount. If a cotton or linen garment has just a few small marks on it, spot-clean it with a little white vinegar or vodka.

With specially hand-printed fabrics, you should hand wash or gentle machine wash the garment in cold water the first three times, then you can usually wash normally. Regarding silk, a lot of labels say dry-clean only, but most of the time you can actually hand wash them in cool water. Avoid dry-cleaning if possible — I never dry-clean anymore. If you have an oil-based stain on a silk garment, don't wash it with water — put talcum powder onto the stain to soak up the oil, then later wash it. You can wash sequinned, beaded and appliquéd silk by hand if you do it gently. Don't dry it in direct sunlight — better to lie it flat.

Knitted clothes like T-shirts, jumpers and jersey dresses should also be dried flat and stored flat in drawers so they don't lose their shape. Woven clothes are better stored on hangers — velvet hangers for silk and wooden ones for cotton and linen. If you're concerned about moths, line your drawers with lavender-scented paper, put lavender bags all over the place and place a piece of cedarwood in your closet. You can also buy cedarwood hangers, which will deter the moths as well as prevent wrinkling or rust marks from those ancient wire hangers you should have tossed out long ago. If you don't know how to repair clothes either by hand or with a sewing machine, take some lessons. The other reason to look after your clothes properly is because they are in better shape to recycle when the time comes. Make sure you retain the label in the garment, and if you've never worn it be careful to retain the tags.

In grim times, we need a splash of colour — here Marni delivers amply (© Neil Gussey & Verve Magazine).

AFTERWORD

The Future of Fashion

As I was writing this book in early 2020, darkness descended on the world in the form of the Covid-19 virus, infecting millions and killing several million people and putting most of the world into both physical and economic lockdown. I always wondered what would happen to my work if I fell over and had a stroke or something. Now I know. When the virus hit, I, along with a lot of other people, lost much of my livelihood. So, as I was sitting in isolation day after day, drinking gin day after day, writing this book, I came to understand that life as we knew it would never exist again, and that many industries, including fashion and travel, would undergo profound transformation. As I write this, it is 1 December 2020 and the news has just come through that Philip Green's fast-fashion Arcadia Group

(Topshop, Miss Selfridge, etc.) has just gone into receivership. They blamed Covid for the collapse. The next day Debenhams went under.

Mary Portas, the English retail consultant and broadcaster, otherwise known as the guru of the high street, said, 'Covid didn't kill Topshop: a lack of creativity did. In its heyday, Topshop wasn't just a "shop". It was a destination where you could tap into the zeitgeist. Its finger was not simply *on* the pulse — it *generated* the pulse.' As it turns out, Arcadia Group has long faced allegations of labour abuses and using unsustainable practices. Mary's solution is to change things profoundly — she thinks fast fashion is utterly OVER and it was failing way before Covid, maybe 10 years before. The solution is a complete change in values, meaning the introduction of what she calls 'the kindness economy'. This is when an effort is made to make shopping streets more experimental, experiential and diverse, for instance with escape rooms, entertainment, community hubs, sustainable and socially responsible clothing shops, free parking, multi-purpose rental spaces (like pop-ups where one month it could be a frock shop, the next month a meditation space, etc.).

The public have had a think and they now want decency in business and a strong connection between what's in the shop and what's online, i.e. the fashion shops will be smaller, not like grotesque warehouses, and most of the stock will be online — almost like the shop is just a viewing platform with fantastic personal service and consultation. And online platforms like Amazon should pay A LOT MORE tax because that's where everyone is really buying.

The fashion industry was hit very hard by Covid. One of the first fashion houses to announce changes was Gucci. The brand was created in 1921 by Guccio Gucci in Florence, and it made luxury leather goods and bags. They branched into loafers and belts and eventually clothing. In 1994 they hired Tom Ford as head designer. Having completely turned the company around and made it glamorous again, he left in 2004. The current progressive creative director of Gucci, Alessandro Michele, has slashed the frequency of their fashion shows to just two a year, and they will be seasonless, combining women's and mens wear.

Normally a designer would show four times (for the four seasons) a year internationally, maybe even five if you count cruise wear, and it is an excessive, over-the-top, expensive exercise. Michele has said that enough is enough, no more waste, that it is completely unnecessary to adhere to the stale concept of producing endless new clothes, and the whole industry should be seasonless anyway. Clothes should have a long life, shouldn't be thrown away, and making new designs only twice a year gives more time for really creative thought, never mind reducing the huge pressure.

In an Instagram post from lockdown in Rome on 29 March 2020, Alessandro Michele wrote — in slightly poetic, dramatic language: 'These days of confinement, in a suspended time that we can hardly imagine as free, I try to ask myself what is the meaning of my actions . . . We turned out to be so small. A miracle of nothing. Above all we understood we went way too far. Our reckless actions have burned the house we live in. We conceived of ourselves as separated from nature, we felt cunning and almighty . . . We incited Prometheus and buried Pan . . . So much outrageous greed made us lose the harmony and the care, the connection and the belonging . . . At the end of the day we were out of breath.'

The new collection follows Gucci's carbon-neutral commitment, announced the following September, which supports REDD+ projects that protect forests and biodiversity in vital places around the world. Gucci donated €1 million each to the United Nations Foundation's Covid-19 Solidarity Response Fund and Italy's National Civil Protection Department, while producing surgical masks and medical overalls in its factories.

Gucci isn't alone with this new direction. The head designer of Yves Saint Laurent, Anthony Vaccarello, has said he would take note of the waves of radical change being brought about by the concern for sustainability and the slightly right/wrong world of over-consumption. 'It's time to leave the fashion calendar behind,' he said. In an open letter to the fashion industry, the Belgian designer Dries Van Noten asked for a serious re-evaluation of its practices. In it he said (and it was signed by lots of other designers) that this is 'an opportunity for a fundamental and welcome . . . change that will simplify our businesses, making them more environmentally and socially sustainable and ultimately align them more closely with customers' needs . . . Working together, we hope . . . [to] bring back the magic and creativity that has made fashion such an important part of our world.'

There were small positive changes in my own little Covid bubble — my neighbours were very good to me, giving me jam and eggs, and we got to know each other much better. I had a large backyard, which I never used, so gave the use of it to their kids: they set up a sports stadium, outdoor school and hut-building venue. We could have done all these things ages ago in our other lives, but we didn't. On my neighbourhood walks, everybody said hello, having not said one word in the past 10 years. Suddenly privacy and aloofness seemed inappropriate. My friends and I started cooking and delivering meals to each other. I began making inspiring videos for my YouTube channel — on cooking, fashion and stories. One would hope that in the future liberty and democracy will prevail, populism and fear of the foreign will subside, and kindness and positive change will be the order of the day.

Let's hope all this turmoil will be a positive game-changer for the clothing industry; in the meantime we can all do our bit to be sustainable fashion icons.

ACKNOWLEDGEMENTS

As you will have gathered, I owe a huge debt to my mother, but also to my siblings, who put up with being mentioned in my books, and especially to Keriann whose beautiful stitching is featured here.

Thanks to all the designers, crafts people and shops who have indulged me over the years and who are mentioned in these pages. Special thanks to Jane Avery for contributing her story to the fur section.

Thank you to the various photographers whose work is displayed in this book, be they the professionals or the random visitors into whose hands I have thrust a camera and demanded they snap me in my latest find from my shed. Thank you also to the magazines who have given permission for some of these images to be used and for featuring yours truly over the years.

Thank you to Greta Bannister and Aloïs Guinut for the fashion tours, and to all my other fashion friends who have let me pick their brains.

Thanks to Harriet Allan for patience and help in research. Publishers are like mothers — they save you from yourself.